T0134197

The Tears of Re

The Tears of Re

Beekeeping in Ancient Egypt

GENE KRITSKY

OXFORD
UNIVERSITY PRESS

Oxford University Press is a department of the University of Oxford.
It furthers the University's objective of excellence in research, scholarship,
and education by publishing worldwide.

Oxford New York
Auckland Cape Town Dar es Salaam Hong Kong Karachi
Kuala Lumpur Madrid Melbourne Mexico City Nairobi
New Delhi Shanghai Taipei Toronto

With offices in
Argentina Austria Brazil Chile Czech Republic France Greece
Guatemala Hungary Italy Japan Poland Portugal Singapore
South Korea Switzerland Thailand Turkey Ukraine Vietnam

Published in the United States of America by
Oxford University Press
198 Madison Avenue, New York, NY 10016

Library of Congress Cataloging-in-Publication Data
Kritsky, Gene, author.
The tears of Re : beekeeping in ancient Egypt / Gene Kritsky
p. cm.
Includes bibliographical references and index.
ISBN 978–0–19–936138–0
1. Bee culture—Egypt—History. 2. Honey—Egypt—History.
3. Bee products—Egypt—History. I. Title. II. Title: Beekeeping in ancient Egypt.
SF531.E3K75 2015
638'.10962—dc23

2015020031

This book is dedicated to the memory of Dr. Eva Crane.

CONTENTS

ANCIENT EGYPTIAN CHRONOLOGY [WITH MAJOR BEEKEEPING DISCOVERIES]

Neolithic Period (5500–3100 BCE)
Dynasty "0" (3150–3050 BCE)
Early Dynastic Period
  First Dynasty (c.3050–2850 BCE) [first honey bee hieroglyphs]
  Second Dynasty (c.2850–2687 BCE)
Old Kingdom
  Third Dynasty (c.2687–2649 BCE)
  Fourth Dynasty (c.2649–2513 BCE)
  Fifth Dynasty (c.2513–2374 BCE) [Newoserre Any relief, Unas relief]
  Sixth Dynasty (c.2374–2191 BCE)
First Intermediate Period
  Seventh Dynasty (c. 2190–2165 BCE)
  Eighth Dynasty (c. 2190–2165 BCE)
  Ninth Dynasty (c. 2165–2040 BCE)
  Tenth Dynasty (c. 2165–2040 BCE)
  Early Eleventh Dynasty (2134–2061 BCE)
Middle Kingdom
  Late Eleventh Dynasty (2061–1991 BCE)
  Twelfth Dynasty (1991–1786 BCE) [Senwosret III relief]
  Thirteen and Fourteenth Dynasties (1786–1664 BCE)
Second Intermediate Period—Hyksos Occupation
  Fifteenth Dynasty (c.1664–1555 BCE)
  Sixteenth Dynasty (c.1665–1600 BCE)
  Seventeenth Dynasty (c.1600–1569 BCE)
New Kingdom
  Eighteenth Dynasty (c.1569–1315 BCE) [Amenhotep relief, Rekhmire painting]

Nineteenth Dynasty (1315–1201 BCE)
Twentieth Dynasty (c.1200–1081 BCE)
Third Intermediate Period
Twenty-first Dynasty (1081–931 BCE)
Twenty-second Dynasty (931–725 BCE)
Twenty-third Dynasty (c.813–711 BCE)
Twenty-fourth Dynasty (724–711 BCE)
Twenty-fifth Dynasty (c.755–656 BCE) [Harwa cenotaph]
Late Period
Twenty-sixth Dynasty—Saite Dynasty (664–525 BCE) [Pabasa and Ankhhor tombs]
Twenty-seventh Dynasty—First Persian Occupation (525–405 BCE)
Twenty-eighth Dynasty(405–399 BCE)
Twenty-ninth Dynasty (399–380 BCE)
Thirtieth Dynasty (380–343 BCE)
Thirty-first Dynasty—Second Persian Occupation (343–333 BCE)
Greco-Roman Period
Macedonian Dynasty (332–305 BCE)
Ptolemaic Era (305–31 BCE) [Edfu and Dendera honey offerings]
Roman Era (30 BCE–337 CE)

The Ancient Egyptian Chronology was adapted from Clayton (1994), Brewer and Teeter (2007), Redford (2001), and Wilkinson (2008).

PREFACE

As a teenager, I happened upon some honeycomb that had fallen out of a tree near my home in south Florida. I collected the larvae and pupae from the sealed brood cells, placed them into test tubes, and watched them develop into mature bees—an incredible metamorphosis that inspired my lifelong interest in bees and apiculture. My introduction to Egypt began just a few months before I discovered that comb, when I read a book on evolution. The book discussed Bishop Ussher's chronology of the Bible; he calculated that the world was created in 4004 BCE, and that Noah's flood occurred in 2348 BCE. That seemed rather recent to me, so I began to explore ancient Egyptian history to see how it fit with Ussher's chronology. It turned out that the pyramids were built before Ussher's date for the flood, which seemed, to say the least, a bit surprising. To make sure that was true, I started reading a variety of books on Egyptian history and soon was captivated by Egyptian art, mythology, and technology.

My Egyptophilia was reinforced when the Tutankhamun exhibit toured the United States while I was an entomology graduate student at the University of Illinois. Between classes and research, I viewed the exhibit at the Field Museum in Chicago and attended several presentations at the University of Illinois Art Department on the history of Egyptology, and I made a promise to myself that I would go to Egypt within 10 years.

That promise came to fruition in only five years, when I received a Fulbright Scholarship to teach entomology at Minya University in Upper Egypt for the academic year of 1981–1982. During my year in Egypt, I visited 94 archaeological sites and even got locked inside an Egyptian tomb during a sandstorm.

In addition to exploring Egypt's tombs and temples, I studied the role of insects, especially beetles, flies, and bees, in ancient Egyptian culture. After my return to the United States, I read Dr. Eva Crane's (1983) wonderful book *The Archaeology of Beekeeping*, which opened my mind to the field of apicultural archaeology. My wife, Jessee, and I had the good fortune to visit Dr. Crane and her assistant Penny Walker at Dr. Crane's home at Gerrards Cross in Buckinghamshire, England, where we discussed our mutual interests in beekeeping history.

In 2006, I created a travel course to Egypt with my close friend Daniel Mader, an art history professor at Mount St. Joseph University. During our four trips to Egypt with students in tow, I arranged to visit several beekeeping-related sites, where I had the opportunity to make detailed examinations of many important reliefs. In addition to our travel to Egypt, Jessee and I embarked on a sabbatical tour of Europe and the United States in 2011, visiting museums that house important Egyptian collections. These endeavors were crucial in enabling me to complete this project.

A number of colleagues have assisted me in completing this project. Adela Oppenheim at the Metropolitan Museum of Art generously shared her discoveries with me and offered valuable insights that aided in interpreting some of the reliefs. Alice Stevenson arranged for me to examine potsherds and portraits from the Petrie Museum of Egyptian Archaeology at the University College London. David Smart at the Cleveland Museum of Art enabled me to examine the museum's incredible collection of beeswax artifacts, and Elizabeth Saluk kindly provided the photographs of those objects.

Several individuals at the British Museum also offered indispensable help. I thank Daniel Antoine, Marie Vandenbeusch, and Marcel Marée in the Department of Ancient Egypt and Sudan for their assistance in examining and photographing items that were not on public display.

Several people granted permission to publish study photographs or provided photographs to be included in this book. I thank Klaus Finneiser, curator of the Egyptian Museum and Papyrus Collection at the State Museum in Berlin, Audran Labrousse of the Institut Français d'Archéologie Orientale, Francesco Tiradritti, Assistant Professor at the Kore University of Enna and Director of the Italian Archaeological Mission to Luxor, Amihai Mazar from Hebrew University of Jerusalem, Dennis Forbes, the editor of *KMT the Modern Journal of Ancient Egypt*, Thierry Benderitter of www.osirisnet.net, and Beth Cortright for their kindness.

I also thank Peter Der Manuelian, the Philip J. King Professor of Egyptology at Harvard University, who helped me with sources for the hieroglyphic interpretations. George Johnson of *KMT* has been a great help and inspiration for this project. Mohammed Hamdy expedited our travels in Egypt and helped to track down leads when I could not be there myself. Michael Klabunde, Professor of History at Mount St. Joseph University, provided Greek translations and insights to nuances of Egypt's political history. Paul Jenkins, Susan Falgner, Char Gildea, and Julie Flanders of the Mount St. Joseph University Library have tracked down obscure references. My work with the Horticulture

Department at the University of Cincinnati provided resources that expanded this project.

I thank my division dean, Diana Davis, and my colleagues in the Biology Department at Mount St. Joseph University—Elizabeth Murray, Tracy Reed-Kessler, Jill Russell, Meg Riestenberg, Andrew Rasmussen, Heather Christensen, and Richard A. Davis—for their help in attending meetings and meeting with students while I was traveling to examine Egyptian collections. I also thank Mount St. Joseph University for providing a sabbatical and cultivating the academic atmosphere that values this kind of research.

I thank my editor Jeremy Lewis at Oxford University Press for his eye for detail and suggestions that helped to bring this project to completion. I also thank the entire team at Oxford University Press for their support.

This work would not have been completed without the love and help of my wife, Jessee Smith, who traveled to Egypt and Europe with me and kept an eye out for bee-related iconography. Her careful editing of the manuscript and her love of Egypt and the English language have made this project much more enjoyable.

Finally, throughout this book I use the name "Re" instead of "Ra" for the god linked to honey bees in ancient Egypt. Re is the preferred name for the deity in recent Egyptological literature and is pronounced "ray." Similarly, I use "honey bee" rather than "honeybee" for the insect that is the focus of this study. In entomological literature, it is common to split the insect's name into two words because the insect is a true bee. On the other hand, dragonflies are not true flies, so it is written as one word.

The Tears of Re

Introduction

I was exploring the Ramesseum, the mortuary temple for Ramesses II on the West Bank of Luxor, and among the graffiti, I saw the name Belzoni carved deeply into the temple's wall in large letters. I had read about the former circus performer, who was credited with bringing some of the most spectacular antiquities to the British Museum (including the colossal bust of Ramesses II that originally stood in this very temple), but it was another name carved on the wall that intrigued me. Lightly chiseled below Belzoni's name within an incised circle was the name Salt (Figure I.1).

Henry Salt is not as well known as Howard Carter, the discoverer of the tomb of King Tutankhamun, or Belzoni, yet his contributions to Egyptology have been admired by millions of visitors to the British Museum and the Louvre. Salt was the first British consul-general to Egypt, and soon after he arrived in Egypt in 1816, he started collecting antiquities for the British Museum. In fact, it was Salt who employed Belzoni to retrieve the colossal bust of Ramesses II. Salt also funded the excavation of the Great Sphinx, which was buried up to its neck in sand. This excavation revealed the pieces of the Sphinx's beard that can now be seen in the British and Cairo Museums and the Thutmose III stele, which now stands between the Sphinx's front paws. Salt was also responsible for procuring the beautiful paintings from the tomb of Nebamun, which are now displayed on the second floor of the British Museum (Brinton 1979, Manley and Rée 2001, Parkinson 2008).

Of all of the items Salt collected that ended up in the British Museum, a small collection of papyri bearing hieratic writing (the cursive form of hieroglyphs) and dating back to 300 BCE is the inspiration for this book. Included in this collection is a framed papyrus cataloged as the Salt 825 papyrus (Derchain 1965), and it (Figure I.2) includes this fascinating passage:

> The god Re wept and the tears
> from his eyes fell on the ground
> and turned into a bee.

The bee made (his honeycomb)
and busied himself
with the flowers of every plant;
and so wax was made
and also honey
out of the tears of Re.
(Leek 1975)

The god Re was the sun god and the creator of the world, which made him an important deity to the Egyptians. Every day, the rising sun in the east was a symbol of Re's creation, which reinforced his place in their theology. Moreover, this theology was linked with Egyptian politics, and Re was associated with the pharaoh. The pharaohs of the Fifth Dynasty built sun temples in association with their pyramids to further honor Re. Unlike other gods in the Egyptian pantheon, Re did not have a sanctuary that housed his statue. It would have been superfluous, because Re could be seen every day: he was the sun (Müller 2001).

Re's role as creator is the opposite of death, and in Egyptian theology, death is not seen as the end of life but rather its source (Müller 2001). The god Re,

Figure I.1 Graffiti at the Ramesseum showing Salt's carved name. Photograph by Gene Kritsky.

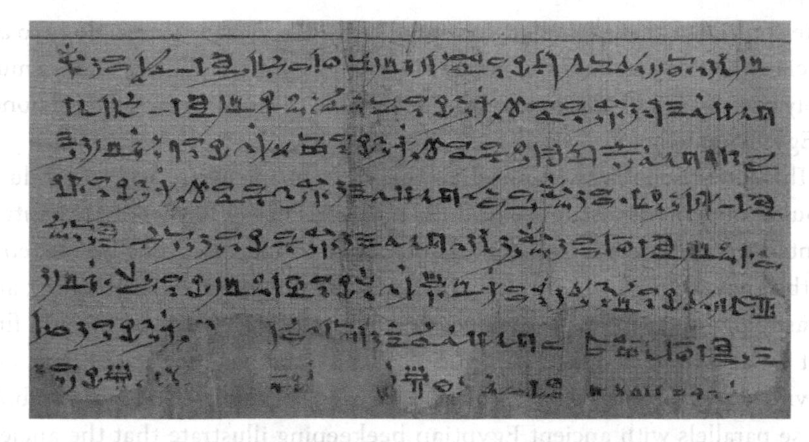

Figure I.2 The section of the Salt 825 papyrus in the British Museum. The story of the tears of Re is told on lines five and six. Photograph by Gene Kritsky. © Trustees of the British Museum.

who was the sun, created honey bees when his tears touched the ground. This view of honey bees as the gifts of Re underscores the importance of bees to the ancient Egyptians.

Appreciating the role that bees and their products have played in Egyptian history requires a cultural immersion. The honey bee hieroglyph can be found in many of the earliest examples of Egyptian writing. During the Old Kingdom, shortly after the construction of the Great Pyramid and the Sphinx, we find carved reliefs showing that the Egyptians had already mastered the art of beekeeping and were processing honey (Kritsky 2010).

Honey was a major commodity that was imported from neighboring kingdoms. It was the primary sweetener in foods. It was a widely used ingredient in medicine, serving not only to make concoctions more palatable but also in treating wounds and cuts. These uses made honey valuable to the extent that it could be used as a means of payment for goods and services.

The Salt 825 papyrus illustrates that bees, as the producers of honey, were important in Egyptian theology, and several gods besides Re were associated with bees. This religious importance required offerings of honey to temples from the Nile Delta region in the north and to the south beyond the great Egyptian city of Thebes.

To appreciate the tears of Re, we must use entomology to document how honey bee biology was applied by the earliest beekeepers. We must become archaeologists to examine the physical evidence for the kinds of hives the Egyptians used. We must use paleography to understand how to interpret hieroglyphs to glean previously lost details about how this ancient apiculture was accomplished. We need to become historians to see how the Egyptian state controlled beekeeping on a national basis. We must use mythology to appreciate the role that bees and honey played in Egyptian religion, and we must apply chemistry and microbiology to determine the medicinal worth of honey in Egyptian medicine.

This interdisciplinary approach to the study of apiculture and its products should not come as a surprise to us. If a researcher 4,000 years in the future wanted to examine our beekeeping, he or she would discover that we relied on the knowledge gained through the fields of entomology, microbiology, and parasitology to keep our bees alive. This hypothetical researcher would find that we also imported honey from other lands, that we produced books and movies that incorporated bees, and that we used bees and beehives as symbols. These parallels with ancient Egyptian beekeeping illustrate that the ancient Egyptian culture, which was among the first to domesticate bees and believed that bees were created from Re's tears, was just as complex as our own.

CHAPTER 1

Beekeeping Begins

The origins of beekeeping are currently lost in the mists of history. The oldest representation of humans interacting with bees comes from a Mesolithic painting found on a cliff in eastern Spain. The painting, which dates back 7,000 to 8,000 years before the present, shows a human hanging from a rope that is dangling from the edge of a cliff, permitting this intrepid opportunist to steal honeycombs from a nest of wild bees (Figure 1.1).

Honey hunting is not the same as beekeeping; essentially, it is searching for a sweet windfall that is there for the taking. In the case of the 8,000-year-old honey hunter in the Spanish painting, it did require more than just finding a colony of wild bees. The hunter had to have the means to reach the comb; the skill, equipment, and fortitude to keep from being overcome by stings; and the assistance of others to help pull him or her up with a payload of honey. Although the skill and courage of early honey hunters might be admirable, it amounts to little more than opportunistic plundering of wild colonies. True beekeeping requires us to provide bees with an artificial cavity in which they can build comb, produce brood, deposit nectar from flowers, and "cook" the honey (Kritsky 2010). Providing a cavity that affords protection to the bees and enables the beekeeper to maintain colonies from one year to the next is part of the symbiosis that exists between bees and humans and forms the foundation of true beekeeping.

Early societies that were capable of initiating beekeeping likely lived in areas with naturally abundant honey bee forage; they had begun living in settled communities, domesticated grain, and mastered pottery or basket making. These pots or baskets would have provided the artificial cavity for the bees to occupy (Crane 1999). It is possible that some observant individual noticed that bees had taken up residence in an empty pot or basket and hoped to encourage more bees to do so by intentionally leaving additional containers out for the bees to occupy. Eventually, containers would be deliberately made for the bees' use, and these were the first true beehives.

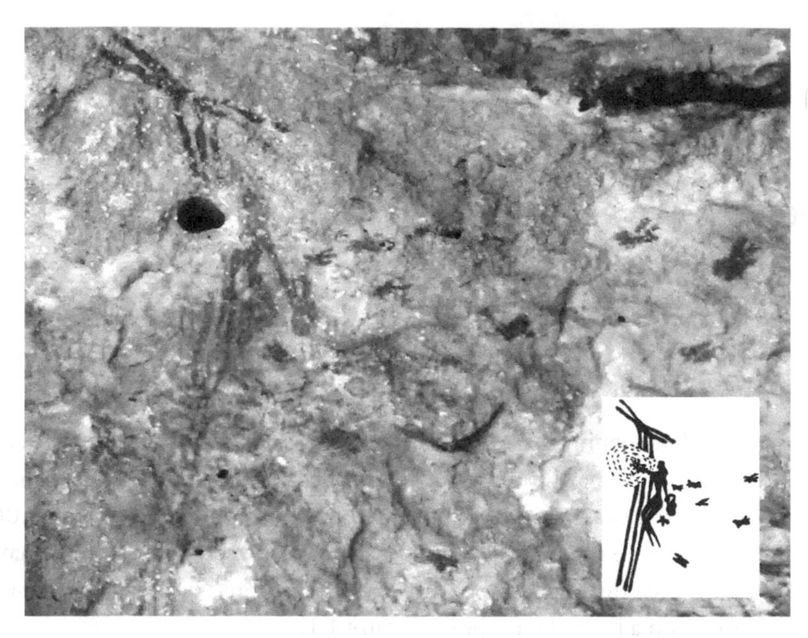

Figure 1.1 The honey-hunting painting from Bicorp, Spain, with a reconstruction shown in the inset. Photo used with permission from www.3wingedfly.com; the inset is from Kritsky (2010) (see color plate 1).

The Fertile Crescent fits the supposed requirements for the development of beekeeping, and there are tantalizing clues suggesting that this area might indeed have been the cradle of apiculture. In 1961, over 400 objects were found wrapped in a straw mat in a cave near the Dead Sea in what is now Israel. Among the items were several copper vessels that appear to have been made using the lost-wax casting process (Department of Ancient Near Eastern Art 2004). This process involves making a beeswax model that was pressed into or encased in sand or clay to create a mold. Firing a clay mold would harden it and melt away the wax. Molten metal (in this case, copper with a certain amount of arsenic) was then poured into the mold, producing a metal version of the original wax model. After the metal cooled, the cast vessel could be lifted out of a sand mold, while a clay mold would be broken away, releasing the cast metal vessel. The straw mat that covered the objects was carbon-14 dated to at least 3500 BCE, making these the oldest known lost-wax cast objects. Some of these items, now on view at the Metropolitan Museum of Art in New York, required beeswax for the production of the original models. The wax clearly came from bees, but this does not necessarily prove that true beekeeping had been established in the region, as beeswax could also have been obtained from comb taken from bees' nests during honey hunting.

At the Louvre in Paris, there are two very large clay cylinders that mention offerings of honey to a temple being built by the Sumerian king Gudea in honor of the god Ningirsu. It had been previously thought that these dated from 2450 BCE, making this the oldest reference to honey in recorded history (Crane 1999), but that date has been revised and the cylinders are now thought to be 300 years younger. However, even if the older date for the cylinders were still valid, the mention of honey does not establish that the Sumerians had developed beekeeping, as honey hunting could also account for these offerings.

The first evidence of an Egyptian connection with bees goes back to nearly 3000 BCE, when the honey bee hieroglyph appeared in the royal titulary to represent the Delta region (Crane 1999). Moreover, Ransome (1986) claimed that there was an official who held the title "Sealer of the Honey" in the First Dynasty. This would imply that either organized honey hunting or true beekeeping was established by this time in Egypt. Newberry (1905) mentioned two titles that related to honey: the "Sealer of the Honey [Jars]" and the "Divine Sealer." Newberry argued that the Sealer of the Honey Jars was among the oldest of the titles in the Egyptian state, and that it related to the "primitive luxuries" that were available to the pharaoh's table. He goes on to state that this title was likely a relic of "the most extreme antiquity."

Clearly, Egyptians 5,000 years ago were aware of honey bees and valued honey, but the use of these titles does not document the practice of beekeeping at this time. This is also true for other early honey bee representations that have been described from Anatolia and central Europe (Crane 1999). Some kind of archaeological evidence of a human-made hive is needed to reveal the origin of true apiculture. Even better would be evidence of hives along with humans processing the hive products for their own use. The oldest such evidence is from the Old Kingdom of ancient Egypt.

CHAPTER 2

The Delight of Re

BEEKEEPING DURING THE OLD KINGDOM

Newoserre Any (Figure 2.1) was the sixth pharaoh of the Fifth Dynasty of the Old Kingdom. He lived a century after the Great Pyramid and the Sphinx had been built on what is now called the Giza plateau, and you can see his pyramid in the distance when looking south from Giza. However, it is not his pyramid that documents ancient beekeeping, but the remains of his solar temple, just over a kilometer to the north of his burial place. In the rubble of that ancient temple, between 1898 and 1901, Ludwig Borchardt (also known for bringing the bust of Nefertiti to Berlin) found the oldest evidence of established beekeeping.

Newoserre Any (c.2474–2444 BCE) called his solar temple Shesepibre, which means "Delight of Re," and it must have been spectacular when it was completed, as it was the largest of all of the solar temples. It was built on a small desert hill close to the fields that were irrigated with water from the Nile. The limited height and uneven ground required that the desert hill be terraced to accommodate the temple. Visitors would enter the temple complex through a small valley temple that had three entrances marked by palmiform columns. This entry connected to a steep causeway that led up the terraced hill to the main temple complex (Figure 2.2a). Once inside the portico, visitors would walk down a short hall that led to corridors on the right and left, and a passageway that continued straight ahead. The corridor to the right led to the temple's storehouse (Figure 2.2b). The passageway would place the visitor in the courtyard with the sun altar in the center (Figure 2.2c). Behind the altar was the focal point of the temple: a spectacular obelisk rising from large pyramidal base. If the visitor proceeded down the left-hand corridor (Figure 2.2d), he or she would eventually reach a small hall at the base of the obelisk (Museum of Reconstructions 2003, Verner 2003). Borchardt called this hall the Chamber of the Seasons (Figure 2.2e) because it was decorated with fine colored reliefs

Figure 2.1 Head and Torso of a King (possibly Newoserre Any), c.2455–2425 BCE Granite, 13 3/8 × 6 3/8 × 5 9/16 in. (34 × 16.2 × 14.1 cm). Brooklyn Museum, Charles Edwin Wilbour Fund, 72.58. Photograph by Gene Kritsky (see color plate 2).

of scenes throughout the year, showing desert animals, birds, men in boats—and beekeeping.

The beekeeping scene from Shesepibre is now displayed at the Neues Museum in Berlin in room 112. It is preserved on five broken limestone blocks that have been reconstructed in two sections. In total, the scene is 61 centimeters long and 16.4 centimeters tall, and it illustrates four beekeeping vignettes that are read from left to right. The carving is a fine bas-relief that was originally painted, but only two of the vignettes still retain some color.

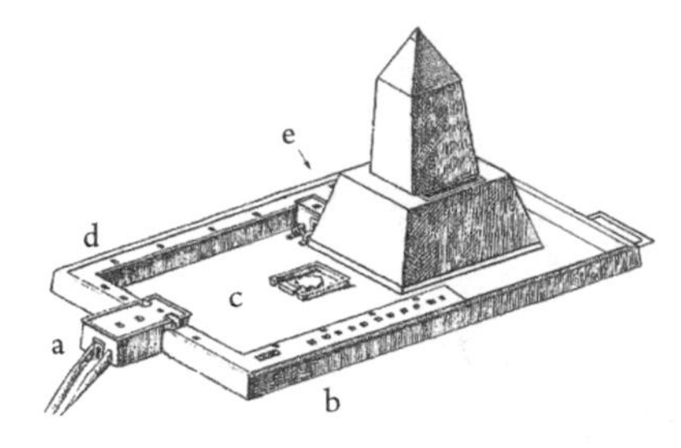

Figure 2.2 A reconstruction of Newoserre Any's solar temple. Modified after Borchardt (1905).

The first vignette (Figure 2.3) is only partially complete. On the far left are the tapered ends of nine horizontal beehives. Only the uppermost hives show the tapering, as the lower hives are represented by just the hive openings. Facing the hives is a kneeling man holding an oval object in his left hand, with his right hand cupped over the end of the object nearest the hives. Above the man are the hieroglyphs N 35 and I 9. The hieroglyph X 1 is to the right of the pair. (These designations are from the Gardiner sign-list, which is a standard reference for identifying hieroglyphs [Gardiner 1988].) The three hieroglyphs are the letters for the Egyptian word *nft*, which Edel (1974) interpreted as "to blow." Crane (1999) reported that Quirke interpreted this as "to create a draught or current of air." Unfortunately, part of the oval object the man is holding is missing, and we cannot see what the man is doing with it, as his face is also missing. It has been suggested that he is using this object like a pottery smoker, but it differs from other smokers depicted in later Egyptian reliefs, which casts doubt on this scenario.

Figure 2.3 The left two vignettes of the beekeeping relief at Newoserre Any's solar temple. Photograph by Gene Kritsky. Publication courtesy of the Egyptian Museum and Papyrus Collection, State Museum, Berlin (see color plate 3).

An alternative explanation of this scene is described by Edel (1974) and Kuény (1950). In their interpretation, the oval object is not a smoker but rather a container used by the beekeeper while he was calling the queen. Crane (1999) described how traditional Egyptian beekeepers call queens to determine whether young queens are present in the hive, indicating the colony is preparing to swarm. If this was the case in the first vignette, the beekeeper was not smoking the bees but actually managing swarming by inducing the queen and workers to enter the jug as he called the queen. This conclusion could be supported by careful examination of the beekeeper's right hand, which is holding the end of the oval container, perhaps trapping the artificial swarm inside. Kuény noted that the hieroglyphic word *nft* also includes the root of the word that means "emitting a breath or a little sound."

In calling the queen, the beekeeper mimics her audible communication, which is called piping. She makes the sound by pushing her thorax against the comb and vibrating her wing muscles without moving her wings. The call consists of a second-long tone followed by a series of shorter bursts (Shimanuki et al. 2007). Root (1888) accurately described the sound of piping as "zeep, zeep, zeep." A young queen will begin piping right after she emerges from her queen cell. If there are other queens developing in the hive, they will answer the newly emerged queen while still in their cells with a call that is often at a different, distinctive pitch. This call has been described as "quacking" (Root and Root 1923). Kuény (1950) described how experienced traditional beekeepers in Egypt place their mouths close to the entrance of the hive and make a "kak, kak" call several times to imitate the call of a newly emerging queen. If there is a mature queen in the hive, she will respond to the beekeeper. If this is what is depicted in the first vignette, then this technique of calling queens has been practiced for over 4,500 years, and the ancient Egyptians' understanding of bee behavior was much more advanced than we might have guessed.

The second vignette (Figure 2.3) is separated from the first by a dividing line at the top of the register. At the far left of the vignette is a standing man facing to the right, and he is emptying the contents of an amphora into a large jug that is resting on the ground. To the right of the large jug is a kneeling man, also facing to the right, who is holding a taller vessel with both hands. On the other side of this tall vessel is a standing man who is holding a small pot equipped with a spout. The man is holding the base of the pot with his left hand and the spout with his right hand, and he is pouring the contents of the pot into the tall vessel.

Above the scene are the hieroglyphs G 17 (an owl), v 23 (a whip), and v 28 (a wick). These are the letters of the word *mḥ*, which means "to fill, fill up, or to pour." Kuény (1950) thought the first man on the left was emptying honey that had been taken from a hive into the large jug, and that the man holding the jar

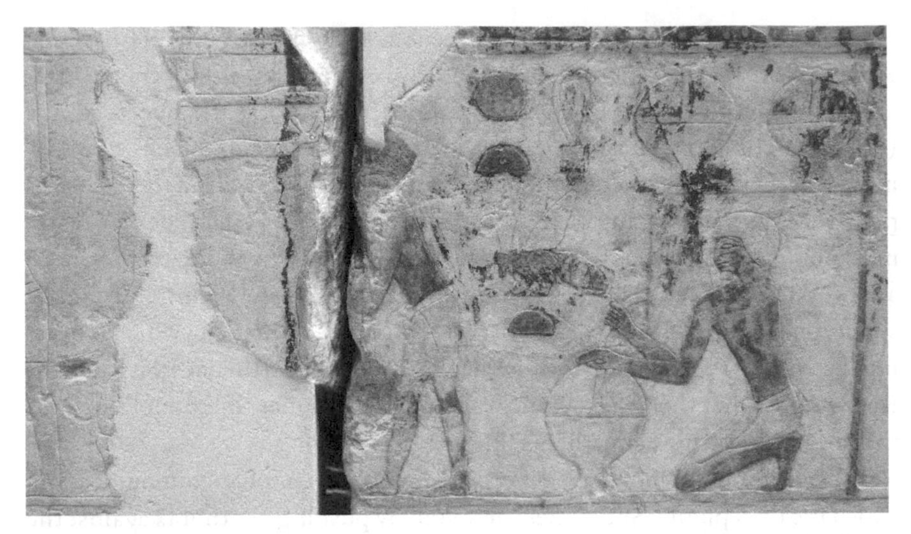

Figure 2.4 The right two vignettes of the beekeeping relief at Newoserre Any's solar temple. Photograph by Gene Kritsky. Publication courtesy of the Egyptian Museum and Papyrus Collection, State Museum, Berlin (see color plate 4).

with the spout was pouring water into the taller vessel to dilute the contents. Another interpretation is that the man is pouring honey that has been separated from the wax. The spout on the container is situated on the side of the vessel, and if it was filled with crushed honeycomb, the wax would float to the top, allowing the honey to be decanted through the spout that is below the wax at the surface.

The third vignette is the most damaged of the four (Figure 2.4). Like the previous vignette, this one is separated from the previous scene by a line at the top of the register. On the left of the scene is a kneeling man facing to the right, but all that is preserved is the back of his head, part of the side of his face, the back of his neck, and his upper back. Below the missing portion is the back of his hips and the sole of his foot. On the other side of the missing section is the edge of his right shoulder and forearm, but unfortunately, it is impossible to make out what this man is doing.

A second man is standing facing the kneeling man. His head is missing, but the outside portion of his right arm is preserved and extended downward. Only the outer edge of the left arm from below the shoulder to the elbow is preserved. The orientation of the arms suggest that this man is holding a container similar to the standing man in the second vignette. Presumably there is a vessel between the two men, but this portion of the relief is completely destroyed, making confirmation impossible. Above the man on the left are the hieroglyphs M 17 (the reed leaf), D 36 (an arm with an upward-facing palm), and I 9 (an asp). This is the word *afj*, which means "squeeze" or "press."

Figure 2.5 A line-art representation of the entire beekeeping relief at Newoserre Any's solar temple. Reconstruction by Gene Kritsky.

The fourth vignette is complete, with some of its color preserved (Figure 2.4). Here, a left-facing kneeling man ties a lid onto an almost spherical container with a conical-footed base. Directly above him is a shelf bearing two similar containers that have also been tied shut. Between these containers are three hieroglyphs: the letters Aa 1 (a filled-in circle), S 20 (a cylinder-seal attached to a necklace), and X 1 (a loaf of bread). This is the Egyptian word *ḫtm*, which means "seal."

Below the hieroglyphs for the word "seal" and to the upper left of the container being sealed by the kneeling man are two hieroglyphs: L 2 (a honey bee) and X 1 (a loaf of bread). This is the Egyptian word *bît*, which is the word for "bee." When combined with a vessel, it forms the word "honey." The combined hieroglyphs have been interpreted as the Egyptian word *ḫtm bjjt*, and translated as "sealing honey" (Edel 1974).

The four steps illustrated, in order from left to right (Figure 2.5), are quieting the bees or calling a young queen; pouring honey and possibly diluting the honey; pressing or squeezing the honey, presumably to get rid of particulates; and finally sealing and storing honey. This elaborate process used at least five different types of containers. As this relief was in the Chamber of the Seasons and associated with the return of migratory birds, it is thought that these activities were typical of November or fall labors (Edel 1974).

Kuény (1950) has suggested that part of the scene might be a brewing scene, and that the Egyptians were using honey to brew a form of mead. Ikram (1994) has claimed that there is no evidence that the Egyptians used honey to make wine; however, Sagrillo (2001) has noted that the Egyptians did add honey to beer and wine. If Kuény's view is correct—that this scene illustrates some form of brewing—the man holding the spouted jug in the second vignette might be pouring honey into a wine vessel, or he could be pouring water into a container of honey in order to dilute it. In the Old Kingdom tomb of Mereruka, the container used to capture the liquid extracted from grapes is similar to the first vessel on the left in the second vignette (Wilson 1988). Regardless of the specifics of the relief at Shesepibre, this relief shows that beekeeping and honey processing were held in high regard during the time of Newoserre Any, and that beekeeping had already developed into a sophisticated occupation, clearly indicating that true beekeeping had begun well before the date of this temple.

Little is known about Pharaoh Newoserre Any's reign; even its duration is uncertain. In a room adjacent to the Chamber of the Seasons is a relief of his *sed*-festival, which was celebrated once every 30 years of a king's reign, giving us a minimum time period for the length of his rule. We do know that he presided over considerable tomb and temple construction, which was carried out by a complex administration of officials, including individuals involved with beekeeping and honey.

Egypt was one of the first nations to develop a complex organization to run the affairs of the state. At the top of the political organization was a state administration that dealt with the royal court and defined geographic borders. The development of writing made this system of central government possible, enabling the detailed communication that would be critical in coordinating any national efforts, but distance between settlements required another level of administration that focused on a province or region. At the most local level of organization, there were temple administrations that governed these important religious and economic centers. All three entities could receive tributes to cover the cost of their operations (Haring 2001, Pardey 2001, and Quirke 2001). The fact that there was an official Sealer of Honey during the First Dynasty implies that honey was being stored and accounted for, presumably because of its value, and it may have constituted part of such a tribute.

The individuals carved in relief in the first beekeeping scene appear to be nameless workers, but during the time of Newoserre Any, there was an individual who apparently had the responsibility of monitoring the harvesting of honey. His name was Nykara.

Nykara's tomb was discovered by C. M. Firth at Saqqara and likely mentioned in his report for his 1925–1926 field season. The tomb finds included a number of statues (Figure 2.6) and part of the tomb's false door (Scott 1952), an architectural element of the tomb often placed in a western wall. The door functioned as the connection between the living and the dead and was where offerings were placed for the deceased (Shaw and Nicholson 1995, Wiebach-Koepke 2001). Nykara's false door is on display in the Cleveland Museum of Art (Figure 2.7). Only nine blocks survive: one forms the drum (the top of the inner section), and four blocks each form the left and right door jamb. The innermost section below the drum is missing (Berman 1999).

Nykara is seen on both jambs, facing inward toward the missing central section and holding a staff. The inscriptions on the drum reads, "The greatest of

Figure 2.6 Seated statue of Nykara, 2408–2377 BCE. Egypt, Old Kingdom, Fifth Dynasty, reign of Newoserre Any or later, 2408–2377 BCE. Red granite and pigment; 53.4 × 20.5 × 28.0 cm. The Cleveland Museum of Art, Leonard C. Hanna, Jr. Fund 1964.90 (see color plate 5).

the tens of Upper Egypt, Nykara." On the left jamb is written, "The greatest of the tens of Upper Egypt, scribe, overseer of the granary of the residence, overseer of the waterfowl, overseer of the palace, overseer of Hut-ihut [city in the western Delta] [overseer of] book-scribes of the granary, overseer of the houses of the royal children in the two houses, priest of [Sah]u[re], priest of Re

Figure 2.7 False door of Nykara, 2408–2341 BCE. Egypt, Old Kingdom, Fifth Dynasty, reign of Newoserre Any or slightly later, 2408–2377 BCE. Limestone; 168.0 × 111.5 cm. The Cleveland Museum of Art, Leonard C. Hanna, Jr. Fund 1964.91.

in Shesepibre, priest of Set[ib]ta[wy], Nykara." His titles continue on the right jamb: "The greatest of tens of Upper Egypt, scribe [overseer of] the two granaries, overseer of all marshlands, [overseer of] officials, overseer of all hunters, overseer of all beekeepers, overseer of all sunshade-bearers, chief of the granary, priest of Re in [the temple of Neferirkara?] [Set]ib[re], priest of Horus, Nakara" (Berman 1999: 130–131, García 2010). The statement that Nykara was "priest of Re in Shesepibre" places him during or just after the reign of Newoserre Any, as Shesepibre was the name of Newoserre Any's solar temple, where the beekeeping relief was located.

The statues of Nykara that have been found help complete our impressions of this important Fifth Dynasty administrator. At the Brooklyn Museum is a painted limestone statue of Nykara seated with his wife standing to his left and his son to his right. A red granite statue of a seated Nykara is in the collection of the Cleveland Museum of Art. The Metropolitan Museum of Art has a painted limestone statue of Nykara seated between his wife and his daughter, as well as a statue of Nykara sitting cross-legged in the manner of a scribe. For the first time, we can place a name and a face with someone involved with ancient Egyptian beekeeping, although it is unlikely that Nykara, a wealthy family man, was ever directly involved with beekeeping; rather, he was the highest-ranking individual to oversee the honey harvests.

The importance of bees can also be seen in the Fifth Dynasty during the reign of Unas, who became pharaoh approximately 40 years after Newossere Any's death. Although Unas did not construct a solar temple, he did have a small pyramid built at Saqqara (Figure 2.8) (Clayton 1994), which was connected to his mortuary temple by a long causeway (Figure 2.9). Selim Bey Hassan (Figure 2.10), recognized for his excavations at Giza, excavated the Causeway of Unas in 1937 and 1938. The tomb of King Tutankhamun had been discovered only 16 years earlier and interest in Egyptology was still

Figure 2.8 The pyramid of the pharaoh Unas at Saqqara. Photograph by Gene Kritsky.

Figure 2.9 The causeway connecting the pharaoh Unas's valley temple, and his pyramid as it appears today. Photograph by Gene Kritsky.

Figure 2.10 Selim Bey Hassan in Giza between 1934 and 1939. Photograph from Matson Collection, Library of Congress Prints and Photographs Division.

strong, so it is not surprising that Selim Bey Hassan's discoveries at the Causeway became global news, as shown from an Associated Press account in the *New York Times* published on July 10, 1938. The headline teased, "City of 20,000 Dead Found at Sakkara [*sic*]," indicating that the dig uncovered "at least 20,000 mummies." The necropolis in question was, in reality, destroyed by Unas during the construction of his causeway (Altenmüller 2001). The Associated Press report went on to describe the discoveries at the causeway, including reliefs that "showed work in the fields of the four seasons, including gathering of honey and figs" ("City of 20,000 Dead" 1938).

Selim Bey Hassan (1938) revealed that the causeway was 666 meters long and 6.7 meters wide, and that it was constructed of Tura limestone. The ceiling (of which only a small section is now restored) was covered with stars painted on a blue background. Reliefs on the walls documented several scenes and activities: the importation from Aswan of the granite used in Unas's pyramid, metal-working, Unas with several gods, and activities of the seasons, including the "collection of honey." Hassan (1938: 520) continued, "Some of these scenes occur in the Sun-temple of [Newossere Any] at Abu Gurab." (Abu Gurab is the modern name for the site of the solar temple.) Two years later, William Stevenson Smith (1940: 148) reiterated the similarity with the solar temple in his report on excavations in Egypt. He wrote, "For the Fifth Dynasty, the causeway of the Unas temple, excavated by Selim Bey Hassan in 1937–1938, added a very valuable series of scenes to the known repertoire of the temple decorator. Among these was a rare parallel to the 'seasons' scenes of the Abu-Gurob Sun Temple of [Newossere Any]."

Unfortunately, Hassan did not include an illustration of the honey-collecting relief in his report. Instead, the artifact was likely placed in storage in one of the magazines (unadorned tombs used to store archaeological finds) at Saqqara. Further study was halted by World War II, during which archaeological work in Egypt essentially stopped.

The tantalizing hints of the honey-collecting scene continued to appear in the literature. In 1957, Ludwig Keimer (1957: 25) wrote about the Unas honey relief, "A small fragment of a similar scene was discovered early during the last war among the reliefs which decorated the causeway that led up to the pyramid of Unas at Saqqara (Fifth Dynasty, about 2350 B.C.)." Crane (1999: 164) added, "During the excavation of the Causeway of Unas (c. 2350 BC) in the 1940s (*Chronique d'Egypte*, 1938), a relief was found which also showed activities through the seasons, with a small fragment of a beekeeping scene similar to that in [Figure 2.5]. It was mentioned by Keimer (1957), but no description has been published."

Apicultural interest in the Unas relief passed unnoticed by Egyptologists, and the relief remained in storage for decades—that is, until Audran Labrousse and Ahmed Moussa of the Institut Français d'Archéologie Orientale completed a detailed study of the reliefs found at the causeway of Unas and published their findings in 2002 (Labrousse and Moussa 2002). They observed that the honey-collection scene is part of a badly broken portion of the causeway. The scene (Figures 2.11–2.12) shows some similarity to the fourth vignette of the relief from Newoserre Any's solar temple. At the bottom right

Figure 2.11 The fragments that include the Unas honey-collection relief. Published with permission of Audran Labrousse.

Figure 2.12 A line drawing of the fragments that include the Unas honey-collection relief. Published with permission of Audran Labrousse.

of the preserved section are three sealed jars similar to the round sealed jars in the Newoserre Any relief; the very top of a fourth jar appears to the right of the three. Above the jars are the hieroglyphs s 29 (a folded cloth), four N 33 (grains of sand), S 16 (a faience necklace), L 2 (a honey bee), and the back of a bird. The authors have translated the hieroglyphs as "a hekat of honey." (A hekat is an ancient Egyptian unit of volume of approximately 4.8 liters.)

It is not surprising that Hassan or others who may have seen the honey-collection relief did not go to the trouble to describe it in detail, as it did not include any new information about beekeeping as such but was rather a straightforward depiction of a collection of honey. However, the fragment does reinforce the importance of honey collection and, by extension, beekeeping during the Old Kingdom.

The death of Pharaoh Unas marked the end of the Fifth Dynasty, but not the end of beekeeping. As the Sixth Dynasty progressed, honey production had increased to the level that honey became a valuable trade commodity. The evidence for this is found at the tomb of Sabni, who was the governor of Elephantine in Aswan during the reign of Pharaoh Pepy II. The tomb is reached by a felucca sail across the Nile, and then a steep hike up to the rock-cut tomb. Sabni engaged in a number of trade expeditions to Nubia, including one to recover the body of his father, who died or was killed on a similar expedition. According to the biographical account on the tomb wall, Sabni took 100 asses carrying oil and honey. He may even have taken beeswax, as the inscription mentions a kind of oil from an unknown source (Kadish 1966, Sowada 2009).

CHAPTER 3

Instability and Reunification

BEEKEEPING DURING THE MIDDLE KINGDOM

The Old Kingdom ended shortly after the death of the Sixth Dynasty king Pepy II, who was the longest-reigning pharaoh in Egyptian history. His death did not herald the beginning of the Middle Kingdom but eventually led to what historians call the First Intermediate Period, also called Egypt's "dark period" because so little is known about this time (Clayton 1994). It is beyond the purview of this book to recount all of ancient Egyptian history, and even Egyptologists disagree about the details of what actually happened at this time. What is generally accepted is that following the death of Pepy II, the central authority of the state disintegrated, and provincial administrations rose to greater importance. Contributing to this decline is the possibility of global climate change that may have resulted in widespread famine in Egypt. The Seventh and Eighth Dynasties comprised leaders who claimed legitimacy from Memphis, the capital of the Old Kingdom. The Ninth and Tenth Dynasty leaders were centered in Herakleopolis, near modern-day Beni Suef, and they controlled Middle Egypt. The overlap of these two centers of government in time is reflected in the four dynasties that overlap in the chronology presented at the beginning of this book (Clayton 1994, Franke 2001).

The major leaders of the Eleventh Dynasty were centered in Thebes in Upper Egypt. Nebhepetre Montuhotep from Thebes defeated the rulers from Herakleopolis and reunified the country, initiating what historians call the Middle Kingdom. When standing on the upper levels and looking towards the east, visitors to Deir el-Bahri, the great temple of Queen Hatshepsut, will see the ruined remains of Nebhepetre Montuhotep's temple just to the south.

The lack of beekeeping and references to honey during the First Intermediate Period is not surprising, as few records survived this time of turmoil. However, Middle Kingdom documents show that beekeeping remained alive and well in Egypt. The central administration went through a time of instability and breakdown during the First Intermediate Period. Provincial administrations, therefore, stepped in to fill the administrative void, and the occupations that thrived during the Old Kingdom continued during this time.

The first evidence of Middle Kingdom beekeeping comes from the time of Senwosret III, who was the fifth king of the Twelfth Dynasty and is known for his military accomplishments. Visitors to the British Museum, the Louvre, and the Metropolitan Museum of Art will quickly recognize Senwosret III—his statues show him with heavy eyelids and a down-turned mouth that gives him a dour expression. He conducted four southern campaigns and eventually extended Egypt's boundary beyond Nubia. He also increased the authority of the state administration and reduced the power of the provincial administrations.

Senwosret III constructed his pyramid at Dahshur, approximately 3.6 kilometers northeast of the Bent Pyramid of the Fourth Dynasty king Snefru. The Egyptian Expedition of the Metropolitan Museum of Art excavated part of the causeway associated with Senwosret III's pyramid and uncovered blocks with a few scenes reminiscent of the Chamber of the Seasons, including a partial beekeeping relief (Oppenheim 2011).

The relief shows two sets of four stacked horizontal hives, and like the hives at Newoserre Any's solar temple, these hives have tapered ends. Between the two sets of hives are three bees facing alternately right, left, right from top to bottom. The relief does not add to our knowledge of how beekeeping was practiced during the Middle Kingdom, but it does show that the hive design remained the same as during the Old Kingdom (Oppenheim 2013).

Egyptian beekeeping continued to be organized during the Middle Kingdom, and the evidence comes from a very small source: a scarab seal. Seals were used from the Early Dynastic period, primarily as a method to keep containers intact. The earliest seals found in Egypt were cylinder seals: tube-shaped objects with incised carvings. Strings were threaded through the tube, permitting them to be worn around the neck.

By the Sixth Dynasty, seal-amulets were being used. These were circular, square, or oval with at least one flat surface, upon which was carved the design. A seal-amulet from the Sixth Dynasty at the Cleveland Museum of Art features a carved honey bee (Figure 3.1). Some of these seal-amulets were carved in the form of a variety of animals, including frogs, crocodiles, or the head of a hawk. Scarabs were soon included in the amulet menagerie, and by the end of the First Intermediate Period, scarab seals were the predominant form.

The scarab seal that contributes to our understanding of Middle Kingdom beekeeping dates from the Thirteenth Dynasty and is in the British Museum. It is carved of steatite, also known as soapstone, and is only 2.4 centimeters long, 1.7 centimeters wide, and 1.1 centimeters thick. It is carved with two lines indicating the prothorax and the wings, and additional lines

Figure 3.1 Seal amulet, 2311–2140 BCE. Egypt, Old Kingdom, Sixth to Eighth Dynasties, 2311–2140 BCE. Bone; 2.6 cm. The Cleveland Museum of Art, Gift of the John Huntington Art and Polytechnic Trust 1914.684.

indicate the spiny legs. The underside is margined with scrolls that surround the carved hieroglyphs, which translate to "Chief Beekeeper, King's Acquaintance" (Figure 3.2) (British Museum 2013, Martin 1971). This official title of chief beekeeper or overseer of the beekeepers suggests that the title Nykara held during the Fifth Dynasty continued during the Middle Kingdom.

One object from the Middle Kingdom, occasionally cited as a possible beekeeping artifact, deserves consideration. It was found by W. Flinders Petrie between 1888 and 1890 at Illahun, near the opening to the Faiyum Oasis. Illahun, often called Lahun or Kahun, was the Twelfth Dynasty location of the pyramid of Senwosret II, the father of Senwosret III. Near the pyramid was the town that Senwosret II built to support the people employed to administer his burial cult. Petrie recognized that the town would be a "prize" filled with historically dated objects (Tyldesley 2005).

Petrie's 1890 volume *Kahun, Gurob, and Hawara* included illustrations of the burial goods, amulets, and pottery that were discovered during the digs. Petrie wrote that among the more "peculiar objects" were "long pipe-like" pots that were closed at one end (Figure 3.3). One of these long pots, now in the Manchester Museum, was examined by Harold Inglesent, who was a chemist and a beekeeper. Inglesent was interested in determining whether any of the pots in the museum's impressive collection might have contained

Figure 3.2 The Middle Kingdom scarab, 30550, with the inscription Chief Beekeeper, King's Acquaintance. Photograph by Gene Kritsky. © Trustees of the British Museum.

honey. In this long pot, he found evidence of beeswax, some pollen grains, and the metatarsus of a bee (the tip of the hind leg). His conclusion was that this long pot was an ancient Egyptian beehive (David 1986). Unfortunately, the pot is too small to support the hypothesis that it was hive; it is only 38 centimeters long, 9 centimeters wide at one end, and 7 centimeters wide at the other (Crane 1999). This is considerably smaller than the horizontal beehives still used in traditional Egyptian beekeeping and other ancient hives from the Middle East. Crane (1999) suggested that this might have been a model of a beehive. There are many examples of models of everyday activities, but this pot is much larger than those that would be typically found in Egyptian models. A simpler explanation for the presence of beeswax might be that this pot contained collected honey.

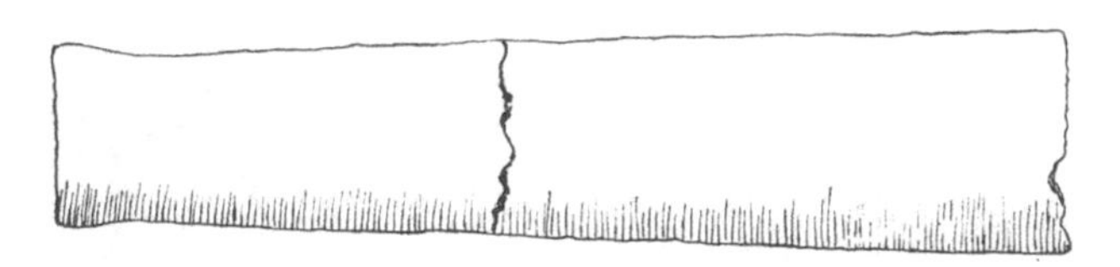

Figure 3.3 The long tubular vessel that was thought to be a model beehive. From Petrie (1890).

Toward the end of the Thirteenth Dynasty, the political unity that defines the historical designation of kingdoms was broken with the contemporaneous rise of the Fourteenth Dynasty kings who reigned in the eastern Delta for approximately 57 years. This period of political instability permitted a foreign group, the Hyksos, to invade Egypt from the eastern desert, marking the beginning of the Fifteenth Dynasty and the Second Intermediate Period.

As the Hyksos were reigning in Egypt primarily in the north, kings of the Seventeenth Dynasty rose in Thebes. These kings controlled much of southern Egypt and maintained the essence of the Egyptian culture of the Middle Kingdom. Late in the Seventeenth Dynasty, the Theban pharaoh Ahmose was able to push the Hyksos out of Egypt and reunite the country, launching the greatest period of Egyptian history—the New Kingdom (Clayton 1994, Quirke 2001a).

Honey retained its importance as a valuable commodity during the Second Intermediate Period. Kamose, the last pharaoh of the Seventeenth Dynasty and the brother of Ahmose, started the resistance against the Hyksos, and he recorded his successes on two steles at Karnak. The first stele is badly damaged, but fortunately the details are preserved on the Carnarvon Tablet (Gardiner 1916: 16–17). It recounts the successful plundering of the Hyksos city of Avaris in the Delta: "My soldiers were as lions are with their spoil, having serfs, cattle, milk, fat and honey, dividing up their property, their hearts gay." The second stele, which is now in the Luxor Museum, continues, "I haven't left a plank to the hundreds of ships of fresh cedar which were filled with gold, lapis, silver, turquoise, bronze axes without number, over and above the moringa-oil, incense, fat, honey, willow, box-wood, sticks and all their fine woods—all the fine products of Retenu [Palestine]—I have confiscated all of it!"

CHAPTER 4

The Age of Empire

BEEKEEPING DURING THE NEW KINGDOM

The New Kingdom, which lasted nearly 500 years, was the pinnacle of Egyptian influence, artistic development, and stability. It started with the Eighteenth Dynasty, which included some of the most important figures in Egyptian history: Queen Hatshepsut, whose successful reign prompted the noted Chicago Egyptologist James Breasted (1916: 83) to call her the "first great woman in history"; Thutmose III, who extended Egypt's boundaries beyond Syria in the north and south into Nubia; Akhenaten, who broke with the established theology of ancient Egypt and founded the cult of Aten (the first experiment of monotheism); Nefertiti, Akhenaten's wife, known to us through her stunning bust now in Berlin's Neues Museum, who may have ruled Egypt on her own following Akhenaten's death; and Tutankhamun, the "boy king" whose intact tomb hints at the unknown splendor of the other tombs of the great pharaohs.

The riches of the Eighteenth Dynasty included several tombs and other discoveries that expanded our understanding of Egyptian beekeeping. The oldest of these tombs is that of Amenhotep (TT73) (TT73 stands for Theban tomb 73, the 73rd non-royal tomb discovered on the west side of the Nile from Thebes). When the tomb was described in 1957 by Säve-Söderbergh, the owner's identity was unknown, but it was apparent that the individual was an important official during the reign of Hatshepsut. Eventually, the tomb's owner was identified as Amenhotep, the overseer of the cattle of Amun, the superintendent of the work of the two large obelisks of Amun, and chief steward and warrior of the king.

Amenhotep's tomb includes a badly damaged beekeeping scene on the subregister of the northeast wall. Säve-Söderbergh (1957) interpreted the objects on the right of the relief as a stack of beehives, and he confirmed his identification by reconstructing the damaged parts of the relief, which established that the objects were indeed beehives. The hives are stacked in a manner similar to those in the relief from Newoserre Any's solar temple and have similar blunt ends.

The highly fragmentary scene does show some noteworthy details that help complete a reconstruction (Figure 4.1). First, the two beekeepers are doing different things. The back of the figure at the top is nearly straight, whereas the figure at the bottom is leaning in toward the hive. The upper beekeeper is holding something that has a sharper corner than the object held by the lower man. Säve-Söderbergh suspected that the sharper corner was part of a censer, a clay bowl that was used to hold incense or some kind of fuel to produce smoke, as seen in greater detail in a later tomb. The more rounded object held by the bottom figure could be honeycomb or some kind of pot. Unfortunately, not enough detail is preserved to allow a precise identification. Whatever the object is, the bottom beekeeper is holding it directly in front of the opening of a hive.

This scene is significant as it is possibly the oldest depiction of using a censer as an offering of incense to the bees. If incense was used as an offering, the

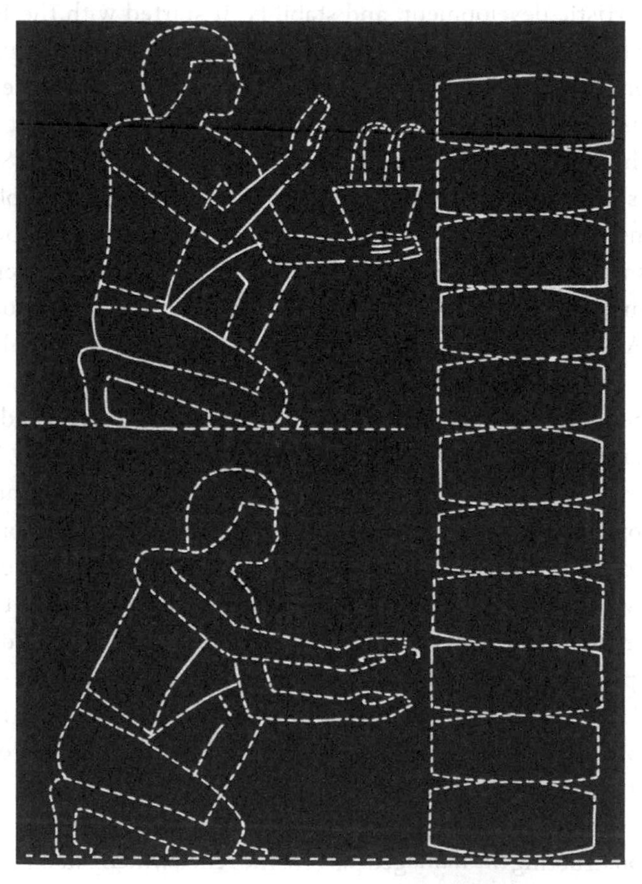

Figure 4.1 A reconstruction of the badly damaged beekeeping relief at the tomb of Amenhotep (TT73). Modified from Säve-Söderbergh (1957).

smoke may have quieted the bees, giving the impression that these important insects appreciated the gift. As such, the Egyptian beekeepers may have discovered the technique of smoking the bees to calm them as an accidental side effect of religious ritual.

The vizier Rekhmire lived approximately 30 years after Amenhotep, during the reign of Thutmose III. A vizier was the highest-ranking official of Egypt, comparable to a prime minister or chief of state in today's world. The vizier administered justice, managed royal estates, controlled irrigation, and supervised both royal and private security. The vizier would also have been the chief authority over the beekeepers. In Rekhmire's day, the office of vizier was inherited, and Rekhmire's grandfather, his uncle, and possibly his father held the position. During Rekhmire's term in office, he served two pharaohs, Thutmose III and Amenhotep II.

Rekhmire's beautifully painted tomb (TT100) is at Qurna, a private cemetery on the West Bank of Luxor across the road and up a short hill from the Ramesseum, the great mortuary temple of Ramesses II. Like the pharaohs' tombs in the Valley of the Kings, it was cut directly into the rock. The area in front of the door forms a courtyard, the back wall of which frames the undecorated entrance to the tomb (Figure 4.2). The tomb has been known for centuries, having been investigated by Robert Hay in 1832 and eventually fully excavated in 1889 (Davies 1943). The entrance to the tomb passes through a

Figure 4.2 The entrance to the tomb of Rekhmire in the Valley of the Nobles. Photograph by Gene Kritsky.

Figure 4.3 The beekeeping painting in the tomb of Rekhmire. Photograph by Gene Kritsky (see color plate 6).

short vestibule that opens up into a large transverse room extending to the left and to the right. Directly ahead and through a slightly constricted opening is a long corridor that increases in height as it goes deeper into the tomb and back toward the false door.

The painted interior of the tomb includes numerous examples of Rekhmire receiving honey as a payment of taxes on the right side of the transverse room. Straight ahead, along the ceiling at the top of the left-hand wall, is the most complete beekeeping scene in Egypt. The scene is crowded into a long narrow triangle, a composition that was forced on the painter by the vaulted ceiling. Painted along the top of the scene is a kheker frieze, a series of bundled reeds that resemble the tied tassels of a carpet (Figure 4.3), and storehouses filled with goods are painted in the register below the scene.

The beehives (Figure 4.4) are shown at the far right of the scene as three horizontal oblong structures, flat on the left side with rounded ends on the right. Unlike the earlier scenes, these hives are resting on a platform. Their gray color suggests that they may be made of dried mud. The rounded ends have been interpreted as the front of the hives, and the men are working them from the rear (Crane 1999). One man is standing and holding a censer that

Figure 4.4 Smoking and removing honeycombs from the hives, as seen in the tomb of Rekhmire. Photograph by Gene Kritsky.

is producing smoke, either to quiet the bees or as an offering to them. Next to him is a kneeling man who is removing a section of rounded white honeycomb and placing it into bowls that are resting on the platform. Behind the man holding the censer and toward the top of the scene is a large red bowl with a pronounced rim, overflowing with crushed honeycomb.

In the drawings published by Davies (1943), there is a small bee drawn within the space between the two beekeepers' arms. However, I was unable to find the bee when I made a detailed examination of the tomb. Indeed, an earlier illustration by Newberry (1938) does not include the bee; he wrote that Davies had later informed him of a faint bee, which Davies included in his reconstruction. It may be too faint to be seen from ground level today, or it might have been lost when the tomb's paintings were cleaned.

Left of the two beekeepers working the hives and the large bowl is a man with his back to the beekeepers, holding a large jar by the top (Figure 4.5). Behind it are four additional jars, the right sides of which can be faintly seen below his arms. Directly above the jars is a large flat-bottomed object with a round top. Its color matches the color of the tops of the jars the man is holding. Facing the standing man is another man sitting on a stool, working with similar but shorter jars. Above these jars are three dark yellow pitchers; the handle of the foremost pitcher can be faintly seen on the left side. Like the jars, the right edges of the other two pitchers can barely be discerned. This scene may depict the separation of the wax from the honey, which is eventually poured from the pitchers into another container.

Behind the seated man is a squatting man holding the edge of a diamond-shaped container with his left hand (Figure 4.6). In his right hand, he is holding a small white cup. Facing him is another man holding the lid of the diamond-shaped container. Directly behind this man are four of the diamond-shaped

Figure 4.5 Detail of honey processing; from the tomb of Rekhmire. Photograph by Gene Kritsky.

Figure 4.6 Sealing honey in diamond-shaped containers; from the tomb of Rekhmire. Photograph by Gene Kritsky.

containers, the first two stacked on each other and the remaining two placed singly on the ground.

These diamond-shaped containers are made up of two shallow bowls, with the top bowl held upside down over the bottom bowl. Presumably, the bowls were sealed together with beeswax. These containers were the standard illustration for a set quantity of honey; they also appear in the tax collection paintings in the transverse room.

The beekeeping scene is the final scene of a much larger tableau, which starts farther to the right of the beekeeping scene. Rekhmire is seated facing to the left, dressed in his vizier clothing and holding a staff in his right hand and the Kherp scepter in his left (Figure 4.7). The hieroglyphic inscription above

Figure 4.7 A schematic showing Rekhmire overseeing the delivery of goods and the processing of grain. From Newberry (1900).

him reads, "receives *uah*-grain and honey in the treasury of the temple, and seals every precious thing in [the house of Amun], in his office of Overseer of the Secrets." Behind him are 12 scribes.

In the top register facing Rekhmire are a scribe and three servants who are measuring the grain that was delivered to them in baskets and poured into a heap. The middle register shows Rekhmire facing men who are grinding the grain to make flour (Figure 4.7). The text reads, "pounding the *uah*-grain in the treasury of Amun in order to make loaves." To their left (Figure 4.8) are men called "slaves of the Department of Dates," separating the fine-grained flour from the coarser flour. To their left are bakers who are mixing the flour with honey and fashioning it into wedge-shaped loaves, which are then baked in the brick ovens. The four-handled amphora along the top of the scene is labeled "honey."

The bottom register shows the "tribute of all countries" being delivered by a dozen "captains of the boats" (Figure 4.7). These "captains" are the men bowing to Rekhmire. The tribute is being carried to the left to fill the buildings below the beekeeping scene (Figures 4.8 and 4.9). The items being delivered include sandals, mats, amphorae, shields, ebony, ivory, ostrich feathers, gold, silver, copper, malachite, incense, olive oil, and cinnamon bark.

Visitors to the tomb will find the beekeeping scene difficult to see clearly, as it is in the front of the room up against the poorly lit ceiling, and it has suffered loss of detail over the years. A reconstruction of the beekeeping relief

Figure 4.8 A schematic showing the delivery of goods (bottom register) and the making of honey cakes (top register, far left). From Newberry (1900).

Figure 4.9 A schematic showing Rekhmire's beekeeping scene above storage building. From Newberry (1900).

(Figure 4.10) illustrates what the original scene might have looked like. Two items stand out in the scene. First, the hives are different from any of the hives we have encountered to date. Their rounded ends and flat openings are very different from the tapered hives in the tomb of Amenhotep, which was built just a few decades earlier. The hives are also resting on a platform, whereas the earlier hives were on the ground. Second, this is the first certain depiction of the use of smoke with bees, and it is consistent with the sharp-cornered, bowl-shaped censer held by one of Amenhotep's beekeepers (Figure 4.1). In fact, it would not have been possible to accurately identify the object in Amenhotep's tomb if Rekhmire's beekeeper was not using a similarly sharp-cornered censer.

Almost contemporaneous with Rekhmire's tomb and also at Qurna is the tomb of Nakht (TT52). Nakht was called a "scribe" and the "Observer of the Hours" at Amun's temple. The latter title is sometimes interpreted to mean an astronomer, but in ancient Egypt, it likely indicated that the title holder made certain that specific rituals took place at specified times (Hodel-Hoenes 2000). The tomb was likely discovered in 1889, which contributed to the preservation of the vivid colors of the paintings. It was a favorite destination of

Figure 4.10 A reconstruction of the Rekhmire beekeeping scene based on firsthand observations by the author.

visitors during the early 20th century, and Baedeker (1902: 289) refers to the tomb as "wonderfully brilliant."

Unlike Rekhmire, who ranked right under the pharaoh, Nakht was a middle-level official, and the differences in the tombs' size and decoration reflect this different status. The date of his tomb is unknown, but the style of the paintings suggested to early Egyptologists that he lived during the reign of Amenhotep II or early in that of Thutmose IV. In recent years, some have argued that it dates to late in the reign of Thutmose IV. Although Nakht was not a high-ranking official like Rekhmire, his tomb testifies to the importance of honey in the afterlife of the upper middle class. Visitors enter the tomb from the courtyard through a short passageway into the transverse chamber. On the north side of the east wall of this room is a painting showing Nakht and his wife, Tawy, making an offering of incense to the gods Amun, Re-Horakhty, Osiris, and Hathor. Nakht and Tawy are facing a stack of goods and just right of center, a heron is standing upon a diamond-shaped container of honey (Figure 4.11).

Figure 4.11 The offering table with a container of honey in the tomb of Nakht. From Davies (1917).

This uniquely shaped container is also found in the nearby tomb of Menna (TT69). Menna is thought to have lived during the time of Thutmose IV and was likely a contemporary of Nakht. Menna was the "Scribe of the Fields of the Lord of the Two Lands of Upper and Lower Egypt." In this role, he is thought to have been an archivist or surveyor keeping tabs on land ownership (Hodel-Hoenes 2000).

Menna's tomb is reminiscent of the tomb of Nakht, having a courtyard, an entrance to a short vestibule that connects to a transverse room with an east and west side, and a passage or doorway to a long chamber opposite the entrance. Visitors to this tomb will see several scenes that are similar to those of Nakht, but what stands out are the vividly colored paintings, the result of an outstanding restoration project (Hartwig 2013). The walls of Nakht's tomb are behind glass and poorly lit in comparison, making them seem dingy. On the other hand, Menna's tomb opens eastward, and visitors in the morning get to appreciate the striking colors on a bright white background.

Immediately to the left when entering the transverse room, on the east wall of the entrance is a large painting of a seated Menna receiving goods. He is sitting on a stool that has an animal skin on the seat (Figure 4.12). He is holding a staff in his right hand and folded cloth in his left, which are symbols of his status. The translation of the text above him reads, "Enjoying himself with the

Figure 4.12 The offering table with two containers of honey in the tomb of Menna. Photograph by Gene Kritsky (see color plate 7).

work of the fields, by the great confidant of the Lord of the Two Lands, in [his] wish . . . the eyes of the King in every place, Overseer of the estate [of Amon, Men]na, justified before the great god" (Hirst 2012). Among the gifts at his feet are baskets of fruit, yellow birds, and geese. Above the birds are three goblets and three baskets of fruit, topped with a nest of birds, a nest of eggs, two fish, two cucumbers, and two diamond-shaped containers of honey (Wilson 1988). At the top are three tables laden with grapes, dates, or figs, with lotus flowers resting on top of the fruit.

Entering the long chamber on the right bottom register is a procession of porters bringing offerings including a cow for slaughter, birds (identifiable even though this part of the painting has been damaged), and a man holding his arms outward to each side of his body, balancing a diamond-shaped honey jar in each hand (Figure 4.13).

Rekhmire's tomb was not the first tomb to use this diamond-shaped honey jar. That honor may belong to the Theban tomb TT343 belonging to Benia, which may date to the early Eighteenth Dynasty and the reign of Thutmose III. Benia was an overseer of works or construction projects. His name is Asiatic or possibly Hebrew. His Egyptian name was Pahekamen, which is the most common name used in his tomb. The tomb has the same floor plan as the previously discussed tombs, with an entrance leading to a transverse room and a center passageway to a long chamber.

To the left of the entrance to the transverse room, in the same location where it was found in Menna's tomb, is a painting of Benia accepting cattle,

Figure 4.13 A porter carrying two containers of honey in the tomb of Menna. Photograph by Gene Kritsky (see color plate 8).

geese, and honey. Unlike the scene in Menna's tomb, here the top tier of gifts includes two jars of honey with lotus flowers draped over them. Behind the table are three procession lines bringing goods to the tomb, including a man carrying two diamond-shaped honey jars in the same manner as Menna's procession. If Benia's tomb does date back to the early Eighteenth Dynasty, this would be one of the oldest examples of this unique honey jar (Benderitter 2007).

The diamond-shaped honey jars also occur in the tomb of Userhat (TT56), another Eighteenth Dynasty official. Userhat was an upper-middle-class official serving as "Scribe Who Counts the Bread of Upper and Lower Egypt." He would compare the amount of wheat that was being harvested with the amount of bread being produced. His tomb is built on the typical plan, and honey jars are found to the left on the entrance wall. Userhat dominates the scene, looking at three registers that show cattle being driven and branded. Directly below Userhat are three diamond-shaped honey jars among the harvested food that is being delivered to him.

On the wall opposite the cattle scene is the painting of the "beautiful festival of the valley," a yearly remembrance of the dead. The deceased's relatives would gather at the tomb for a feast, music, dancing, and an offering to the deceased. These events likely took place in the courtyard in front of the tomb. The bottom register includes an offering table on which are three diamond-shaped honey jars (Benderitter 2008).

The diamond-shaped honey jar also occurs in the late Eighteenth Dynasty relief of the Royal Cupbearer Tjawy, on display at the Museum of Fine Arts in Boston. The honey jar is in the center of the fifth register toward the bottom of the relief. This is the only example of an incised honey jar, as all the previous examples are paintings.

While the previous depictions are very similar, the tomb of Tjener (TT101) has a rather different painting of a honey offering. Tjener was the cupbearer of the king, and he lived during the middle Eighteenth Dynasty during the reign of Amenhotep II. His tomb is closed, but a painting of the scene by Nina Davies (1936) (Figure 4.14) shows an offering bearer carrying a bowl of honeycomb in his left hand. The bowl is the same shape as the bottom of a diamond-shaped honey jar, and it contains seven circular combs of honey. The two bees resting on the comb help identify the round objects as honeycomb.

There are other Eighteenth Dynasty paintings of honeycomb. Hugh R. Hopgood painted a basket of honeycomb from the tomb of Qenamun when he was on the Egyptian Expedition of the Metropolitan Museum of Art (2015) during 1914–1916. It is not on display, but it shows two baskets of fruit to the right of a red bowl of honeycomb. Like the honeycomb in the tomb of Tjener, the comb is represented as round sections stacked on end in the bowl, and the

Figure 4.14 The offering of honeycomb from the tomb of Tjener (TT101). Original painting by Nina Davies (1936), and published with permission from the University of Chicago Press.

bowl is the bottom section of the diamond-shaped honey container. There are lotus flowers and buds above the honeycomb, and a bee is painted on an open flower.

Not all honeycomb paintings include the bee. On the second floor of the Egyptian exhibit at the British Museum are the paintings from the tomb of Nebamun that were procured for the museum by Henry Salt. Nebamun was the scribe and grain accountant of the Granary of the Divine, meaning the offerings of Amun. This title places Nebamun in a civil service position for the temple. His funeral offerings include bread, wine, meat, and a bowl of honeycomb. As with the other examples, this painting shows the comb in the bottom half of the diamond-shaped honey jar, just to the left of a plucked bird (Parkinson 2008) (Figure 4.15).

Horemheb, the last pharaoh of the Eighteenth Dynasty, did not have an heir, and his death ushered in the beginning of the Nineteenth Dynasty. The first pharaoh of the dynasty was Ramesses I, who was Horemheb's vizier, and he reigned for only two years before his son Seti I ascended to the throne. Seti I started a major campaign of building and foreign policy initiatives, apparently to help restore

Figure 4.15 The offering of an open container of honeycomb, just to the left of the plucked bird, from the tomb of Nebamun. Photograph by Gene Kritsky. © Trustees of the British Museum.

Egypt, which was still recovering from the reign of Akhenaten. He began work on the Great Hypostyle Hall of Karnak Temple, and he built the great temple to Osiris at Abydos. These structures show Egyptian art at its apogee.

The accomplishments of Seti I were continued and expanded by one of Egypt's greatest pharaohs, Ramesses II. He reigned for just over 60 years, during which there were more monuments built in his honor than for any other pharaoh. In Memphis, just south of Cairo, is the colossal statue of Ramesses II. In Thebes (now Luxor), visitors can examine Luxor Temple, the completed Hypostyle Hall at Karnak, and the Ramesseum. In southernmost Egypt along the Nile, the great temple of Abu Simbel extols the "victory" of Ramesses II over the Hittites at the battle of Kadesh, now part of Syria. Although the battle was a stalemate, Ramesses II accepted a peace offer from the Hittites, and he returned to Egypt claiming victory; he had details of the battle carved at several temples, including the Ramesseum, his temple at Abydos, Karnak, and Luxor, with the largest relief at Abu Simbel (Clayton 1994).

Beekeeping continued to be an important occupation in the Nineteenth Dynasty as documented on steles and temple and tomb walls. Honey is mentioned on Seti I's decree at Nauri. This decree was recorded on a stele that is approximately 35 kilometers due north from the Third Cataract of the Nile in what was considered Nubia. There, two large rocky outcrops rise 90 and 120 meters above the desert, and about 30 meters up the side of the shorter outcrop is the Seti I stele. At nearly five square meters, it is quite large. The decree was issued to consolidate Seti I's economic control over the region by pronouncing the punishments for infringing on the properties of Seti I's temple at Abydos. Gold workers in Nubia were covered by the decree, and those who interfered with the transport of gold to the temple would be severely punished by mutilation and forced labor. The decree also describes the holdings of the Temple at Abydos, including gold, silver, royal linen, incense, and honey "without limit in counting" (Griffith 1927). Such quantities of honey were only possible with large-scale production and controlled distribution.

At the Ramesseum, the mortuary temple for Ramesses II, is a badly damaged stele that lists various officials during his reign (Figure 4.16). The list includes three people with the titles "The Chief of the Stable," an "Administrator of the Palace of Ramesses II," and "*Nefer-hetep*, the Beekeeper (?) of Amon [*sic*]. . ." (Quibell 1898: 20). The Neues Museum's Egyptian collection includes the unfinished stele of Smentu, who was "Head of Beekeepers of His Lord, before Min and Isis." The round-topped stele of Khons at the Kestner Museum in Hanover, Germany, includes his title as "Beekeeper of Amun Great of Victory."

The tomb of another Userhat (TT51) has a very different honey offering scene compared to the earlier examples. Userhat served the cult of Thutmose I but was in office during the reigns of Horemheb, Ramesses I, and Seti I. Userhat apparently died during the time of Seti I, and his tomb was built at Qurna. Its layout is similar to the previous tombs with a courtyard leading to a short portico entrance that opens into the transverse room. Opposite the entrance is a passage to an undecorated chamber that leads to the burial chamber in the rear. To the right of the passageway to the undecorated chamber is a painting showing Userhat and his family worshiping Osiris, who is sitting in a shrine. In front of the shrine is an offering balanced on four narrow stands. The offering includes cuts of meat from sacrificed animals, above which are fruit and honeycombs. However, in this illustration the honeycombs are oval in shape and show a very detailed hexagonal honeycomb pattern (Figure 4.17).

Beekeeping and honey references continue to be found during the Twentieth Dynasty, the last dynasty of the New Kingdom. The Wilbour Papyrus, from the time of Ramesses V, lists various occupations of people in the Faiyum,

Figure 4.16 A schematic of a stele from the Ramesseum, with the title of beekeeper in the bottom register, fourth from the right. Modified from Quibell (1898).

including soldier, stable master, and beekeeper (Emanuel 2013). Visitors to Karnak Temple in Luxor can look for an inscription on the right half of the exterior east wall of the peripteral shrine southeast of the seventh pylon that describes the difficulties of the high priest Amenhotep. On the left portion of the inscription, there is a passage that reads, "It was I who gave barley, emmer, incense, honey, dates, green plants, flowers" (Wente 1966).

Egypt's influence began to decline during the Twentieth Dynasty. The death of Ramesses XI ended the New Kingdom and marked the beginning of the Third Intermediate Period. The first pharaoh of the Twenty-first Dynasty ruled from Tanis in northern Egypt, while a powerful group of priests controlled Thebes. The resulting confusion ended when Pharaoh Shoshenq I (the same Shoshenq, also known as Shishak, who raided Solomon's Temple as described in the Bible) rose to power at the start of the Twenty-second Dynasty.

Figure 4.17 The offering of honeycomb to Osiris, as seen in the tomb of Userhat (TT51). Photograph courtesy of Thierry Benderitter (www.osirisnet.net).

Shoshenq I's campaign of victories in Palestine is recorded on a relief at Karnak Temple, and one of the cities mentioned is Tel Rehov. Tel Rehov, although not in Egypt, is mentioned in numerous Egyptian sources going back to the Eighteenth Dynasty, when the extent of the Egyptian empire included Tel Rehov. During the reign of Seti I in the Nineteenth Dynasty, Tel Rehov remained loyal to Egypt as other cities were rebelling against the Egyptian administration of the region. Seti I recorded this loyalty in a stele erected at Beth-Shean in what is now northern Israel.

Tel Rehov is important to beekeeping history because it is where excavations led by Amihai Mazar from Hebrew University of Jerusalem uncovered the oldest known beehives, dating back to the time of Egypt's Twenty-second Dynasty. They were large cylindrical horizontal hives (Figure 4.18), somewhat reminiscent of those painted on the wall of Rekhmire's tomb. They were approximately 80 centimeters long and 40 centimeters in diameter and were stacked three hives high in parallel rows. The hives were made of baked clay and straw, and a chemical analysis of the inner surface found the residue of beeswax (Mazar and Panitz-Cohen 2007).

It is impossible to document any Egyptian influence on Tel Rehov's beekeeping, so we cannot say for certain if these ancient beehives were exactly like Egyptian hives. What the discoveries at Tel Rehov do indicate is that horizontal beekeeping, as illustrated on the earlier Egyptian reliefs and paintings, was practiced throughout the region.

The history of Egypt during the Third Intermediate Period is a matter of debate, as several dynasties overlap in time. As the Twenty-second Dynasty

Figure 4.18 The beehives found at Tel Rehov. Photo: Amihai Mazar, Tel Rehov Expedition, The Hebrew University of Jerusalem.

progressed, Egypt became politically divided, as documented with the beginning of the contemporaneous Twenty-third Dynasty. During the New Kingdom, the region south of Aswan known as Nubia, the land of the Kush, became powerful enough to form its own kingdom called Napata. Although it was independent, Egyptian theology had taken hold in this region, and the rulers built temples to Amun. With the splintering of Egyptian rule during the Third Intermediate Period, the Nubians moved northward into Egypt in an effort to restore Amun. The Nubian ruler Piya successfully conquered Thebes, initiating the Twenty-fifth Dynasty (Clayton 1994). However, the ruler in Sais in the northern Delta region attempted to limit the Nubian pharaohs' control to Upper Egypt by forming alliances with rulers from several other cities, and thus started the Twenty-fourth Dynasty. Eventually, Piya was able to capture Heliopolis and Memphis, ending the Twenty-fourth Dynasty, which came into existence and ended during the time of the Twenty-fifth Dynasty.

The Assyrians became a major threat to Egypt late in the Twenty-fifth Dynasty, initially by invading Egypt and capturing Memphis. The Nubian pharaoh at the time, Taharqa, retreated southward back to Nubia, and Ashurbanipal, the king of Assyria, eventually sacked Thebes (Clayton 1994). The Twenty-sixth Dynasty, also called the Saite Dynasty, marked the end of the Third Intermediate Period and the start of the Late Period (encompassing the Twenty-sixth through the Thirty-first Dynasties). The Saite Dynasty began when Psamtik I, who was the ruler of Sais, formed an alliance with Assyria, which now controlled Egypt, and he was soon recognized as king. Psamtik I slowly extended his control over the Delta and subsequently moved south to Thebes, pushing out the Nubian pharaohs and forming the last native lineage to rule Egypt. He eventually brought stability to Egypt and became independent of Assyria.

CHAPTER 5

The Saite Dynasty

Beekeeping remained important during the transition between the Twenty-fifth and Twenty-sixth Dynasties, and scenes relating to apiculture continued to adorn tomb walls. One such site belonged to an official named Harwa, who lived toward the end of the Twenty-fifth Dynasty and was the first "Chief Steward of the God's Wife of Amun." This was an important position during the Nubian pharaoh's restoration of the Amun cult, and it meant that Harwa was quite wealthy and influential.

Harwa's badly damaged cenotaph (TT37) is part of the cemetery known as el-Asasif and situated directly across the road from where buses and taxis park at the entrance to Deir el-Bahri, the mortuary temple of Queen Hatshepsut. The cenotaph is being excavated by the Italian Archaeological Mission to Luxor under the direction of Francesco Tiradritti. The cenotaph is closed to the public, but you can look down into its courtyard (Figure 5.1) when visiting other nearby tombs that are open (Naunton 2007).

The courtyard is reached via a portico entrance that connects to a vestibule that opens on the courtyard. A passageway in the back of the courtyard leads to the first of two colonnaded rooms that subsequently reach an Osiris shrine. Excavations have found several shafts; many were likely made well after Harwa's death. His burial chamber is yet to be discovered (Naunton 2007).

What has been found of Harwa's cenotaph indicates that parts of it were copied by two Twenty-sixth Dynasty officials whose tombs include beekeeping-related reliefs. Therefore, it was not surprising when Tiradritti's team uncovered three pieces of a beekeeping relief among the 20,000 blocks collected so far. One piece shows part of a tree (Figure 5.2); a second shows part of a tree and the head, parts of the antennae, and legs of a bee (Figure 5.3); and the third piece shows most of a bee except for the antennae and the tip of the abdomen (Figure 5.4).

The conclusion that these belonged to a beekeeping relief is based on a comparison with the nearby tomb of Pabasa (TT279), who lived during the Twenty-sixth Dynasty, toward the end of the reign of Psamtik I. This tomb

Figure 5.1 The courtyard of the cenotaph of Harwa (TT37). Photograph by Gene Kritsky.

lists his many titles: "Hereditary Prince, True Relation of the King, Beloved of Him; Sole Beloved Friend; Royal Chancellor, Greatest of the Great, Noblest of the Noble; the Mouth of the King which Pacifies the Two Lands, Priest of Amun-Re, King of Gods; Chief of the Priest of the Gods of Upper Egypt; Priest of Monthu, Lord of Thebes; Chief of the Priests of Horus the Great;

Figure 5.2 A portion of a tree from the decoration of the cenotaph of Harwa. Photograph by Giacomo Lovera; © Cultural Association for the Study of Egypt and Sudan NGO.

Figure 5.3 A portion of a tree and a honey bee from the decoration of the cenotaph of Harwa. Photograph by Giacomo Lovera; © Cultural Association for the Study of Egypt and Sudan NGO.

Chief Steward of Amun, and Chief Steward of the God's Wife; and Controller of All the Divine Offices of the Divine Consort of Amun." Psamtik I's daughter, Nitocris, was accepted by the Theban priests as the wife of Amun and essentially became head of the Theban priests. Pabasa, who was the Chief Steward of Amun's wife, would have had priestly and civil responsibilities, suggesting that he was an important individual during the early Saite Dynasty (Lansing 1920).

Pabasa's tomb is also near the parking lot of Deir el-Bahri, and a damaged mud-brick pylon stands near the stairs that lead down to the tomb door that opens to the underground sections. Visitors to the site first enter a small vestibule with a door on the opposite wall to the left. The door opens onto the court of offerings (also called the Lichthof or sun court), a room with columns on both sides of the court and open above, and beyond the court of offerings is the hypostyle hall, with larger four-sided columns on each side. The similarities between Pabasa's tomb and Harwa's cenotaph suggest that parts of Pabasa's tomb were copied from the cenotaph (Nauton 2007).

The beekeeping scene is found in the court of offerings on the south side (facing the entrance to the hypostyle hall) of the second complete column

Figure 5.4 A partial honey bee glyph from the decoration of the cenotaph of Harwa. Photograph by Giacomo Lovera; © Cultural Association for the Study of Egypt and Sudan NGO.

on the left of the entrance to the court. This relief is one of the best known in beekeeping literature, but the images of the relief that are usually published include only the left side of a larger scene (Figure 5.5). Moreover, there is an additional honey-related relief lower on the same column.

The lower left portion of the relief is bordered by sets of two incised lines that form the floor or ground of both the bottom and top panels (Figure 5.6). The outer edge is also marked by two incised lines that are perpendicular to the floor lines. At the far left is a man wearing a kilt and a sash, holding up his hands in the same pose as the hieroglyph A 30, representing worship or adoration. This hieroglyph is also used for words such as "praise," "extol," and "show respect" (Wilkinson 1992).

Right-facing honey bee hieroglyphs (L 2) are found above the man's head and hands. Eight oblong beehives form a stack or wall of hives in front of the man. The ends of the hives are tapered or rounded, similar to the hives in Newoserre Any's sun temple and in Amenhotep's tomb (TT73), but they are not depicted as physically touching each other.

On the other side of the hives are two rows of five bees each, all facing to the right. Numerous bees in such an arrangement are thought to mean "a swarm" (Wilkinson 1992). The left ends of four additional hives can just be seen to the lower right of these bees.

Figure 5.5 The beekeeping relief in the tomb of Pabasa. Photograph by Gene Kritsky (see color plate 9).

Figure 5.6 The lower left section of the beekeeping relief at Pabasa's tomb. Photograph by Gene Kritsky.

Figure 5.7 The upper left section of the beekeeping relief at Pabasa's tomb. Photograph by Gene Kritsky.

Directly above these bees, on the right side of the upper scene, are two rows of seven right-facing bees; four on the upper right side are incomplete (Figure 5.7). Behind the bees, a beekeeper wearing a kilt without a sash is pouring a red liquid into a rounded container similar to the round containers of sealed honey from Newoserre Any's solar temple, but with a less pronounced foot. The Egyptians referred to desert honey as "red" honey, so the color used in the painting may illustrate the honey's origin. Above the man's head are two additional bees, one above the other.

Behind the man are two columns of four objects that are broader at the base and the top compared to the middle (somewhat hourglass-shaped cylinders), with a double line about a quarter of the way down from the top and an inverted V-shape at the bottom. These unusual objects are most likely pot stands or offering stands. Pot stands of a similar design are found in Egyptian reliefs dating back to the Second Dynasty. Shorter pot stands were used with round-bottomed amphorae, but taller pot stands that held smaller, shallower bowls (offering basins) are also known. Visitors to the Metropolitan Museum of Art can see an offering stand from the Fourth Dynasty (catalog number 07.228.24) (Fischer 1973), and another offering stand and basin from the Third Dynasty are in the collection of the Brooklyn Museum (catalog number 34.976)

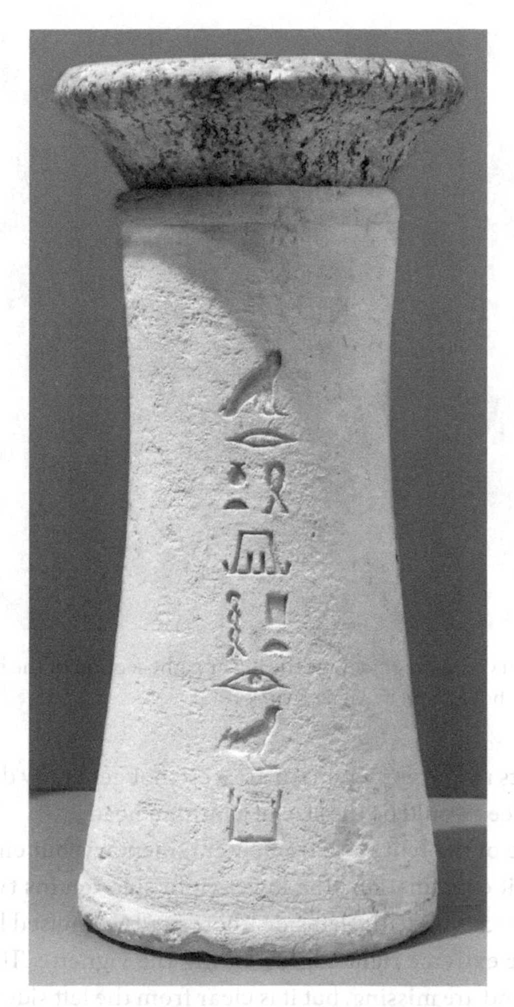

Figure 5.8 Cylindrical Stand with Separate Bowl (Together Forming a Table of Offerings) of the Superintendent of the Granary, Ptahyeruka, ca. 2475–2345 BCE Granite and limestone, 22 3/16 in. (56.3 cm). Brooklyn Museum, Charles Edwin Wilbour Fund, 37.19E. Photograph by Gene Kritsky.

(Figure 5.8). Offering stands have been found dating to the Eighteenth Dynasty at Karnak (Hein 2003) and appear on Nineteenth Dynasty reliefs at Seti I's temple at Abydos. The inverted V at the bottom of the stand is seen in the hieroglyphs W 11 and W 12, which represent stands that hold amphorae (Gardiner 1988).

Tall offering stands like these held shallow bowls for ritual libations. The contents of the libation bowls in Pabasa's tomb are unknown, but the vessels atop the offering stands are the same shape as the container the man in the scene is using to pour honey into the storage jar, although they differ in size.

Figure 5.9 The surviving fragments of the lower right section of the beekeeping relief at Pabasa's tomb. Photograph by Gene Kritsky.

However, objects in ancient Egyptian art were not generally drawn to scale—otherwise, the bees would be the size of a human head.

The right side of the relief is extremely fragmentary, but enough details do remain to permit examination. The lower right side retains two fragments of the scene (Figure 5.9); like the left side, there are two incised lines that form a border along the extreme right edge, framing the vignette. The lines forming the floor or ground are missing, but it is clear from the left side that they would have continued and connected with the lines on the right.

At the far right is a kneeling man facing to the left, wearing a kilt and a sash. The position of his elbows shows that he was holding both arms upward, with the left elbow higher than the right. This is not the same pose as the man on the lower left, whose elbows are at the same height. This would suggest that the left-facing man was doing something different from the man at the opposite side of the relief. Directly in front of the man on the right are two offering stands similar to the stands at the far left of the upper left scene. The second piece of the lower right scene shows the back of the man's head and part of his ear.

The upper right scene is broken into three sections, two of which connect (Figure 5.10). The three sections show that the right border continues into the upper right, and that the floor continues from the left scene. The two connecting pieces show the legs and part of the kilt of a man standing and facing

Figure 5.10 The remaining pieces of the upper right section of the beekeeping relief at Pabasa's tomb. Photograph by Gene Kritsky.

to the left. Directly in front of him are the trunk and lower right branches of a tree. (The presence of this tree in the upper right scene is the primary evidence that suggests that the fragments found in the tomb of Harwa are probably from a very similar beekeeping scene.) The third, separate fragment is a smaller piece showing a man's chin, neck, and upper shoulders. Unfortunately, the man's arms are missing from the relief, so we do not know what he is doing in the scene.

It is possible to partially reconstruct the beekeeping scene using the incomplete figures on the left side of the relief. To the left of the two offering stands in the lower right was a column of eight hives, which would have bisected the entire vignette. The man in the lower right was likely holding something, as the position of the elbows is similar to other reliefs of people holding a variety of objects. The upper register would have included a set of bees facing the tree and a left-facing man on the far right.

A short walk from the tomb of Pabasa along a dirt path to the east leads to another Twenty-sixth Dynasty tomb that belonged to Ankhhor (Figure 5.11), the Steward of the Divine Votress Nitcocris, the Great Mayor of Memphis, Overseer of Upper Egypt in Thebes, and the Overseer of the Priests of Amun. Ankhhor lived approximately twenty years after Pabasa, and he served during the end of the reign of Pasamtik II and the beginning of the reign of Apries. His tomb is not as well preserved as Pabasa's, but what has survived indicates that his tomb was modeled in part after Pabasa's, and it also includes a beekeeping scene.

Figure 5.11 Shawabty of Ankhhor, 595–586 BCE. Egypt, Thebes, Late Period, Twenty-sixth Dynasty, reign of Psammetichus II-Apries. Pale turquoise, vitreous Egyptian blue; 9.9 × 3.4 cm. The Cleveland Museum of Art, Gift of the John Huntington Art and Polytechnic Trust 1914.586.

A long stairway of shallow steps leads to the tomb entrance. To the right is a small room that has a door to the right leading into the Lichthof or sun court, where five square columns still stand. The beekeeping relief is found on the second column on the left as viewed from the entrance. Like the beekeeping relief in Pabasa, the relief in Ankhhor consists of upper and lower registers (Figure 5.12). Unfortunately, three-fourths of the upper register and about a third of the lower register are missing. However, what is present does show several similarities to the relief in Pabasa.

Figure 5.12 The beekeeping relief in the tomb of Ankhhor. Photograph by Gene Kritsky (see color plate 10).

The bottom register is separated from the scenes above and below by two incised lines (Figure 5.13). Unlike Pabasa's tomb, the relief is not framed on the sides by sets of incised lines. At the far left is a kneeling man wearing a kilt and a sash and facing left. His arms are mostly missing, but what is preserved shows that he is holding his hands up with the palms facing forward in praise or worship, as is the man in the similar portion of the Pabasa tomb relief. Between the man's hands and the floor are two right-facing honey bees. The abdomen and wings of a right-facing honey bee are between the man's hands and the floor of the scene above. In front of the two lower bees are seven flattened oblong shapes with rounded ends. These oblongs are distinctly narrower than the hives found in the similar position in Pabasa's tomb. It has been suggested that these are honeycombs rather than hives. The relief above the "honeycombs" is missing; however, there appears to be the very top edge of a honeycomb that would have continued in line with the lower set of honeycombs.

To the right of the honeycombs are two additional right-facing bees, and to the right of the lower bee are the front legs of a left-facing honey bee. The

Figure 5.13 Detail of the bottom register of Ankhhor's beekeeping relief. Photograph by Gene Kritsky.

remainder of the lowest portion of the lower right section is either damaged or missing. The last remaining piece of the lower register has a stack of seven honeycombs, corresponding to the missing honeycombs on the left. To the left of the upper three honeycombs are the ends of a honey bee's hind wing and abdomen. On the other side of the honeycombs is another left-facing honey bee, balancing out the partial bee to the left of the combs.

To the right of the combs are the hands and shoulder of a man holding a *nw* vessel (a nearly spherical pot with a prominent rim) in each hand. This scene is similar to the hieroglyph D 39, the "offering arm," which is often used in a representational setting such the beekeeping relief, and it denotes the offering of a libation (Wilkinson 1992). The two offering vessels represent Upper and Lower Egypt.

The majority of the upper register is missing, which is why most images of this scene show only the badly damaged lower register. However, four pieces of the upper register remain, providing clues as to what the relief probably looked like (Figure 5.14). Three of these pieces are small portions of the floor that were preserved because they were attached to the lower section. At the far left of the upper register, resting on the scene's floor, is the bottom of an offering stand similar to the stands in Pabasa's beekeeping relief (Figure 5.15). The major difference is that there are two columns of offering stands side by side at Pabasa's tomb, while there is only one in Ankhhor's tomb. Immediately to the right of the offering stand are the feet of a right-facing standing man. Next to the feet is what appears to be the lower right edge of a vessel, and to the right of the vessel are the

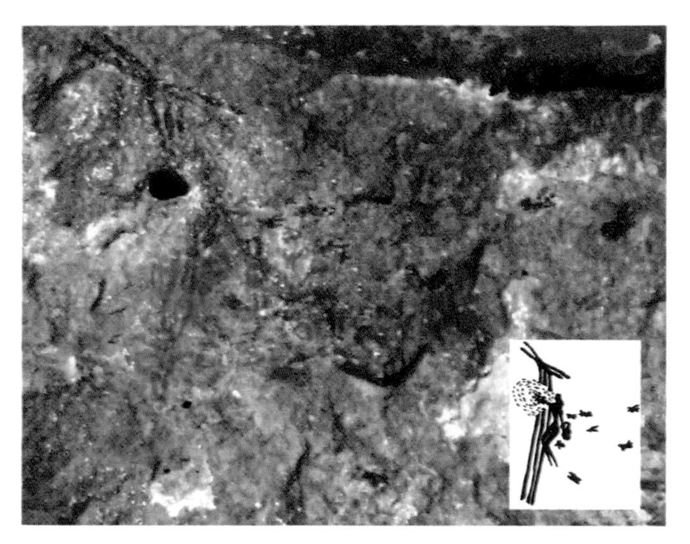

Plate 1. The honey-hunting painting from Bicorp, Spain, with a reconstruction shown in the inset. Photo used with permission from www.3wingedfly.com; the inset is from Kritsky (2010).

Plate 2. Head and Torso of a King (possibly Newoserre Any), ca. 2455–2425 BCE Granite, 13 3/8 × 6 3/8 × 5 9/16 in. (34 × 16.2 × 14.1 cm). Brooklyn Museum, Charles Edwin Wilbour Fund, 72.58. Photograph by Gene Kritsky.

Plate 3. The left two vignettes of the beekeeping relief at Newoserre Any's solar temple. Photograph by Gene Kritsky. Publication courtesy of the Egyptian Museum and Papyrus Collection, State Museum, Berlin.

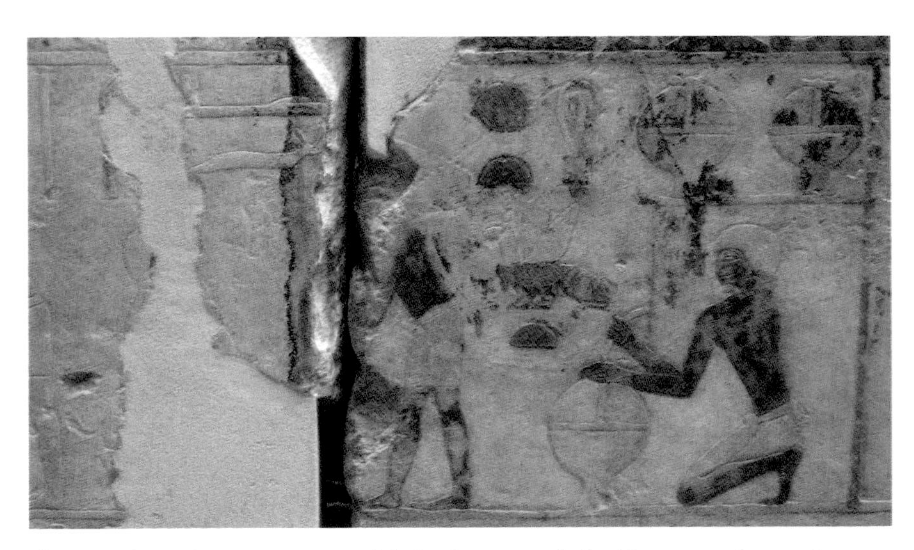

Plate 4. The right two vignettes of the beekeeping relief at Newoserre Any's solar temple. Photograph by Gene Kritsky. Publication courtesy of the Egyptian Museum and Papyrus Collection, State Museum, Berlin.

Plate 5. Seated statue of Nykara, 2408–2377 BCE. Egypt, Old Kingdom, Fifth Dynasty reign of Newoserre Any or later, 2408–2377 BCE. Red granite and pigment; 53.4 × 20.5 × 28.0 cm. The Cleveland Museum of Art, Leonard C. Hanna, Jr. Fund 1964.90.

Plate 6. The beekeeping scene (within blue box) in the tomb of Rekhmire. Photograph by Gene Kritsky.

Plate 7. The offering table with two containers of honey in the tomb of Menna. Photograph by Gene Kritsky.

Plate 8. A porter carrying two containers of honey in the tomb of Menna. Photograph by Gene Kritsky.

Plate 9. The beekeeping relief in the tomb of Pabasa. Photograph by Gene Kritsky.

Plate 10. The beekeeping relief in the tomb of Ankhhor. Photograph by Gene Kritsky.

Plate 11. Details of the upper right section of Ankhhor's beekeeping relief showing parts of two bees, a tree, a man holding his arms in beckoning or summoning, and two hieroglyphs: a filled cup and a twisted wick of flax. Photograph by Gene Kritsky.

Plate 12. Amenhotep III presenting the menu for the Opet Festival as seen in Luxor Temple. The offering of honey is shown in the highlighted box in the center of the top register. Photograph by Gene Kritsky.

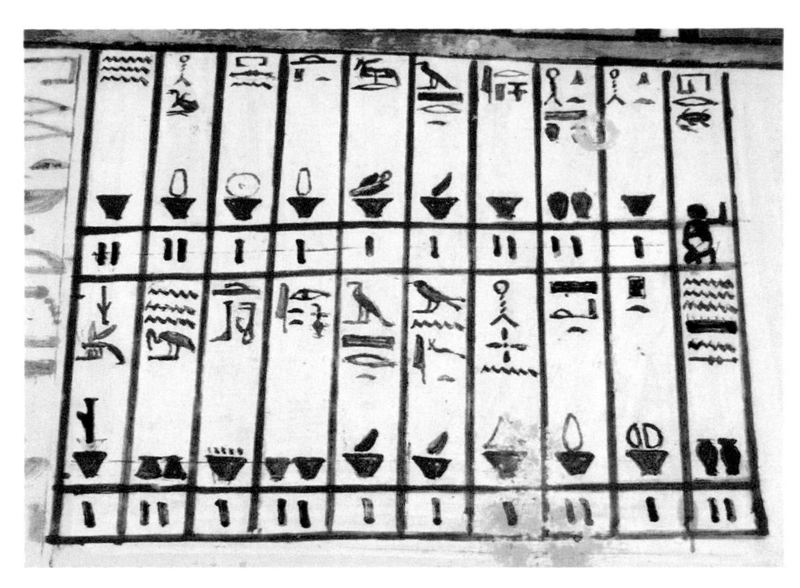

Plate 13. A register listing the food to be provided to Menna in the afterlife. Honey is in the lower far left. Photograph by Gene Kritsky.

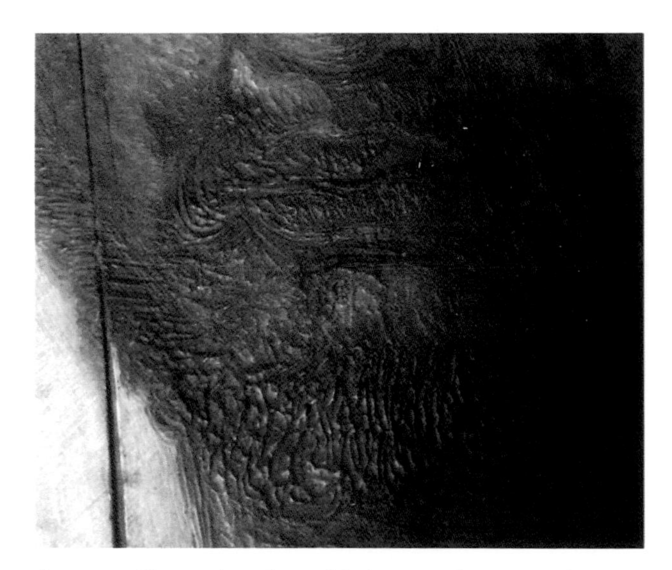

Plate 14. A close-up of "Petrie's Red Youth" showing the textured painting made possible by the encaustic technique, in which pigment was mixed with beeswax. Photograph by Gene Kritsky, and published courtesy of the Petrie Museum of Egyptian Archaeology, UCL.

Plate 15. Four beeswax sons of Horus (from left to right): Hapy, Duamutef, Imsety, and Imsety. Third Intermediate Period, late Twenty-first Dynasty (1069–945 BCE) or early Twenty-second Dynasty (945–715 BCE), The Cleveland Museum of Art.

Plate 16. Four beeswax sons of Horus (from left to right): Hapy, Duamutef, Imsety, and Imsety. Third Intermediate Period, late Twenty-first Dynasty (1069–945 BCE) or early Twenty-second Dynasty (945–715 BCE), The Cleveland Museum of Art.

Figure 5.14 Detail of what remains of the upper register of Ankhhor's beekeeping relief. Photograph by Gene Kritsky.

Figure 5.15 Detail of what remains of the upper left section of Ankhhor's beekeeping relief. Photograph by Gene Kritsky.

abdomen, thorax, and legs of a right-facing honey bee. This is similar to the corresponding scene in Pabasa's tomb, with the man standing in front of offering stands, pouring honey into a vessel, and a line of right-facing bees to the right of the vessel.

Fortunately, the upper right section of the Ankhhor relief (Figure 5.16) includes more detail than the fragments from Pabasa's tomb. At the far right is a left-facing standing man wearing a kilt and a sash. His head and most of his legs are missing, but one left-facing foot is clearly visible on what little remains of the floor. The man's right arm is raised with his palm facing upward and inward, and his left hand is at his side with the palm facing the body (Figure 5.17). This is the same pose as hieroglyph A 26, which is part of the Egyptian word for "call," "summon," or "beckon" (Wilkinson 1992). The hieroglyph for a filled cup (W 10) and a wick of twisted flax (V 28) are carved just above his raised arm.

The man is facing a large tree, of which just the crown and upper branches are preserved, but the narrow trunk of the tree is just to the left of the left-facing

Figure 5.16 Detail of the upper right section of Ankhhor's beekeeping relief. Photograph by Gene Kritsky.

Figure 5.17 Details of the upper right section of Ankhhor's beekeeping relief showing parts of two bees, a tree, a man holding his arms in beckoning or summoning, and two hieroglyphs: a filled cup and a twisted wick of flax. Photograph by Gene Kritsky (see color plate 11).

foot that is found along the bottom of the scene. To the left of the tree are parts of two right-facing bees. The upper bee retains her head, antennae, and the tips of the front legs, and below this bee are the antennae of a second bee. The bees are not in line with the bee from the lower left portion of the register, suggesting that there were two lines of bees in the upper scene, as there are in Pabasa's scene.

Comparing the Pabasa relief with that of Ankhhor, completing the partial images, and filling in the missing and probably symmetrical sections makes it possible to hypothesize what the original Ankhhor relief may have looked like. The bottom relief is framed by the two kneeling men facing each other with the man on the left holding his hands in praise or worship, and the man on the right holding two offerings. There are eight bees or parts of bees found on the relief: five facing right, and three facing left. The available space and symmetry suggests that there were an additional six bees in the lower section, symmetrically arranged into two sets of seven bees facing right and seven bees facing left. Bisecting the two sets of bees would be a stack of fourteen honeycombs or narrow hives running from the top of the register to the bottom. The upper register, from left to right, probably showed three offering stands lined up along the far left with a man pouring honey into a vessel in front of him. There were two sets of four honey bees facing the tree, with a left-facing man summoning the bees toward the tree.

Conversely, we can also use Ankhhor's relief to help reconstruct the missing portions of the right-hand section of Pabasa's scene. First, the kneeling man is probably holding something in offering: likely two *nw* vessels as in Ankhhor's scene. The missing upper section included a tree with a man, who was likely also beckoning bees to the tree.

Unlike the beekeeping scenes from the Newoserre Any's solar temple and the tomb of Rekhmire, the Pabasa and Ankhhor reliefs do not convey any direct information about beekeeping practices, other than pouring honey into a rounded jar for storage. Instead, the position of the hands of the men in the lower left scene, the presence of offering stands, and the kneeling man holding the two *nw* vessels indicate that both reliefs portray a scene of respect or worship. The man summoning or beckoning the bees to the tree may indicate that the Egyptians recognized that bees needed to come to flowering trees to fill themselves with nectar in order to produce honey.

The Saite or Twenty-sixth Dynasty ended when Pharaoh Psamtik III was defeated by the Persians and removed to the Persian capital. The Twenty-seventh Dynasty (the First Persian Occupation) lasted more than a century. It ended with the death of the Persian king Darius II, and Amyrtaeus from Sais declared himself king. He was the only ruler of the Twenty-eighth Dynasty. The Twenty-ninth and Thirtieth Dynasties were also short-lived, but did enjoy some stability, which permitted King Nectanebo I of the Thirtieth Dynasty to rebuild some of the temples. The Thirtieth Dynasty was followed by the Thirty-first Dynasty (the Second Persian Occupation), which lasted only eleven years, before Alexander the Great conquered the Persian Empire ending Egypt's Late Period (Clayton 1994).

CHAPTER 6

The Greco-Roman Period

Alexander, the first of three Macedonian kings, was greeted as a god by the Egyptians. He founded Alexandria, had himself crowned pharaoh, and had his image carved on temple walls in the Egyptian style. Alexander left Egypt to go on to conquer Babylon and extend his rule down to the Indus River. After his death in 323 BCE, Alexander's generals divided up his empire, and Ptolemy became satrap or governor of Egypt. To consolidate his claim to Egypt, Ptolemy kidnapped Alexander's body and brought it to Egypt for burial. He secured his control of Egypt after defeating rival generals and eventually became pharaoh in 305 BCE, launching the start of Egypt's Ptolemaic Era (Clayton 1994, Shaw 2002).

During the Ptolemaic Era, there was a major building campaign that resulted in the construction of several temples, including Edfu and Dendera. Both temples have a few reliefs of honey offerings, confirming that honey production was still highly prized during this period.

The Temple of Edfu is approximately halfway between Luxor and Aswan and is one of the most complete temples to survive to the present day. The temple honors the god Horus, the falcon-headed god who was the son of Osiris and Isis. Construction of the temple began in 237 BCE during the reign of Ptolemy III and was completed in 57 BCE. There are four reliefs at Edfu that include honey offerings, which are easily missed unless visitors make the effort to find them.

The entrance to the temple, flanked by massive falcon statues, is set in a spectacular pylon (Figure 6.1). It opens onto the central court and the entrance to the first hypostyle hall, where the first honey offering is found on the third column of the second row on the east side. The relief shows Ptolemy VIII offering honey to the ram-headed god Mendes Banebdjed. The second relief is found in the inner hypostyle hall (the Festival Hall) on the first column of the second row of columns on the west side. It shows Ptolemy IV Philopater presenting two offering vessels of honey to the god Min, the god of fertility. A third honey-offering scene is found in the Room of Min, which is to the left

Figure 6.1 The Temple of Horus at Edfu. Photograph by Gene Kritsky.

when facing the Sanctuary of Horus. On the top register of the north wall is another scene of Ptolemy IV Philopater presenting two offering vessels of honey to Min (Zecchi 1997). To the right of the sanctuary is a small open court, where the fourth honey relief is on the wall to the right, opposite the steps to the New Year Chapel. This scene shows Ptolemy IV and his wife making offerings to both Hathor and Horus. The impressive offering includes six rows of goods (Figure 6.2), and on the far right of the second row are three vessels of honey (Figure 6.3).

Figure 6.2 Ptolemy IV and his wife's offerings to both Hathor and Horus at Edfu Temple. The honey is highlighted by the box. Photograph by Gene Kritsky.

Figure 6.3 Detail of the honey jars in the offering to Hathor and Horus. Photograph by Gene Kritsky.

The Temple of Hathor at Dendera may include the last honey scene created in ancient Egypt. The temple's construction began around 125 BCE, and its carvings include the famous relief of Cleopatra VII and her son, Caesarion, who was purported to be the son of Julius Caesar. The honey offering is in front of the figures of Cleopatra VII and Caesarion on the outer rear wall on the southwest side of the temple. The deeply carved figures show Cleopatra VII behind Caesarion; both figures are facing to the right, toward six registers of offering goods (Figure 6.4). Two vessels of honey are found just left of center of the third register (Figures 6.5 and 6.6). In this case, the contents of the vessels are identified not by their shape (these containers are tall amphorae), but by the two bees that appear between them.

Cleopatra VII was the last ruler of the Ptolemaic Era. Following the death of Julius Caesar, Octavian and Anthony fought for control of the Roman Empire. A sea battle in 31 BCE gave Octavian the upper hand, and he eventually invaded Egypt, capturing Alexandria the following year. Cleopatra VII, not willing to be taken to Rome in chains, committed suicide, and Egypt became part of the Roman Empire (Shaw and Nicholson 1995).

Roman rule of Egypt changed the administrative control of the country. Emperors were represented in pharaonic dress on temple reliefs, but their involvement was limited to making occasional visits to Egypt; the country was ruled by a prefect who reported to the emperor. Stability in Egypt was maintained by permitting the religion and culture to continue, and temple

Figure 6.4 Cleopatra VII and Caesarion facing six registers of offering goods. Photograph by Gene Kritsky.

construction that was started during the Ptolemaic Era (including Dendera, Kom Ombo, and Philae) was completed during the Roman Era.

Agriculture in Egypt was critical to Rome at this time, as it relied on the great quantities of wheat imported from Egypt on boats sailing from Alexandria. Honey production also continued. Strabo reported that the crocodiles

Figure 6.5 Detail of the offering by Cleopatra VII and Caesarion with the two jars of honey in a box. Photograph by Gene Kritsky.

Figure 6.6 A closer view of the honey offering. Photograph by Gene Kritsky.

at Crocodilopolis were fed meat, grain, and wine mixed with milk and honey. Plutarch wrote that honey was eaten during festivals, and papyri from the fourth century CE also indicate that honey and honey cakes were special festival foods (Frankfurter 1998).

The last hieroglyphic inscription was carved in 394 CE at Philae (Clayton 1994), which many historians mark as the end of the ancient Egyptian culture. Although the ancient empire that built the pyramids, the temples, and the hundreds of tombs adorned with reliefs and hieroglyphic inscriptions had ended, their ancient form of beekeeping in horizontal hives would continue to be practiced in Egypt for nearly two millennia (Kritsky 2010).

CHAPTER 7

The Honey Bee Hieroglyph

The honey bee hieroglyph first appears during the First Dynasty, over two centuries before the pyramids were built at Giza and nearly half a millennium before the first apicultural relief was carved at Newoserre Any's solar temple. These early bee hieroglyphs were not the carefully incised bees that were carved during the height of Egyptian art but were somewhat crude in execution. The earliest glyphs were found by Sir Flinders Petrie (1900) at Abydos, where he excavated royal tombs dating back to the First and Second Dynasties. The glyphs consist of a pointed oval with two antennae at one end and two loops forming the wings jutting from the middle. These early representations were carved with three or even four legs (Figures 7.1a–d and 7.2).

Crude honey bee glyphs like these were soon replaced by a more carefully carved representation of the insect. A Third Dynasty relief at the base of a statue of Djoser at the Imhotep Museum at Saqqara shows the bee hieroglyph carved with a distinct head, two antennae, and a defined thorax with two large wings projecting upward (Figure 7.3). These wings included prominent veins carved along the leading edge. The bee's abdomen is larger and more distinct and is detailed with abdominal segments and a recurved stinger projecting from the tip. Four distinct legs project forward from the thorax, with the hind leg projecting backward and superimposed over the abdomen.

This fundamental design for the bee hieroglyph, established in the Third Dynasty of the Old Kingdom, remained essentially unchanged throughout ancient Egyptian history until the Greco-Roman Period. The bee hieroglyph did not escape the stylistic changes that influenced most Ptolemaic Egyptian art; the bee was depicted in a much more stylized manner, with a more conical abdomen (often out of proportion to the oval thorax) and awkward wings diverging from the base in a V-shape (Figure 7.4a–h).

Honey bee glyphs are often found alongside a sedge (M 23) and two half-circles (X 1) that represented loaves (Figure 7.5). When spoken, the four glyphs would be pronounced "*n-sw-bỉt*," which is translated as "He of the Sedge and the Bee" or "King of Upper and Lower Egypt." The sedge plant represented

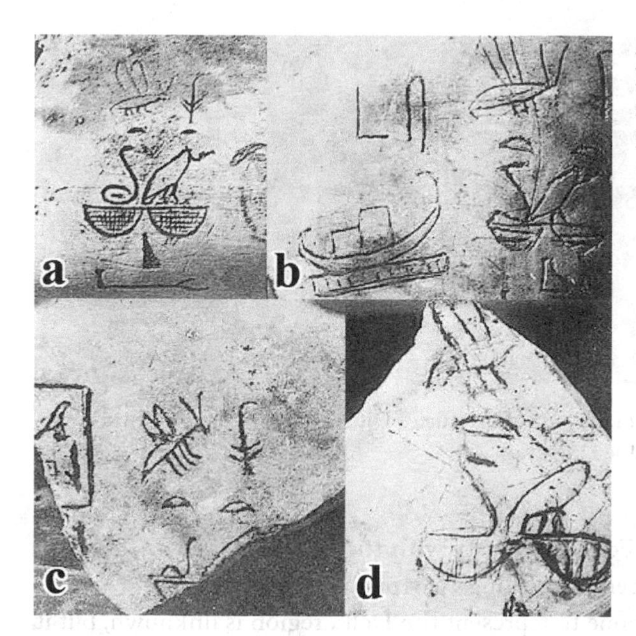

Figure 7.1 Crude early honey bee hieroglyphs from the First and Second Dynasties. From Petrie (1900).

Upper Egypt, and the bee represented the northern Delta region. This phrase was initially used in the First Dynasty, before the practice of carving royal names inside a cartouche began. Crudely carved titularies were found at Abydos, associated with several of the early kings' names (Figure 7.1a–c). Cartouches (from the French word for rifle cartridges, which they resemble) were introduced

Figure 7.2 An ebony jar label from the tomb of the First Dynasty pharaoh Den. The crudely carved honey bee is to the right of the enclosed square in the upper left corner. © Trustees of the British Museum.

Figure 7.3 From the front of the base of a statue of Djoser dating from the Third Dynasty. Photograph by Gene Kritsky.

at the beginning of the Fourth Dynasty with the Pharaoh Snefru. The royal titulary was carved adjacent to the king's cartouche (Figures 7.5).

How the honey bee came to represent the Delta region is unknown, but it has been surmised that the wet, lush Nile Delta would have been prime habitat for honey bees. There is little doubt that the bee hieroglyph represented the

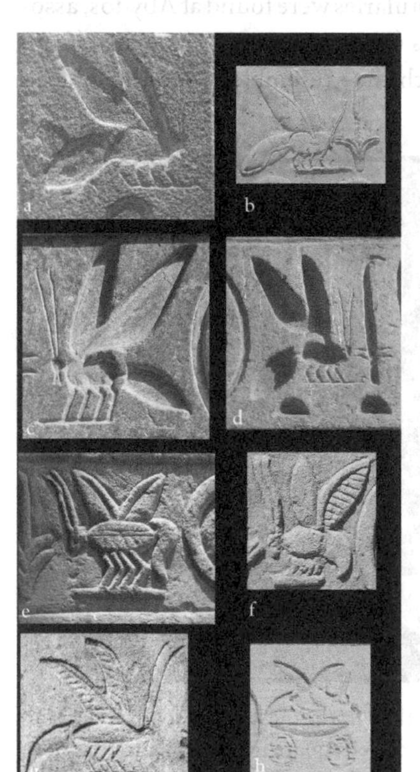

Figure 7.4 The honey bee hieroglyph throughout Egyptian history: a. Tomb of Mereruka, Sixth Dynasty; b. Deir el-Bahri, Eighteenth Dynasty; c. Ramesseum, Nineteenth Dynasty; d. Medinet Habu, Nineteenth Dynasty, e. Kom Ombo, Ptolemaic Period; f. Kom Ombo, Ptolemaic Period; g. Philae, Ptolemaic Period; h. Kalabsha Temple, Roman Era. All photographs by Gene Kritsky.

Figure 7.5 The cartouche of Ramesses III with the royal titulary. Photograph by Gene Kritsky.

Delta, as it was explained in the Kahun papyri from the Middle Kingdom. The passage reads, "He hath come, he hath united the two lands: he hath joined the Reed to the Bee" (Petrie and Griffith 1898: 3).

There has been some question in the past as to what insect was depicted in the bee hieroglyph. Budge (1988) showed the "bee" glyph with the caption "The Bee or Hornet(?)." Petrie also referred to the glyph as "a bee or hornet," but by 1923, he stopped including the word "hornet" (Ransome 1986). The controversy was recently revived when Dodson and Ikram (2008: 98) wrote, "Some scholars regard bees as being symbolic of kingship, although the insect that appears in the *nesu-bity* title was more probably the wasp that is often found hovering near papyrus plants."

It must be kept in mind that the glyph was not intended to be a taxonomically accurate representation of the species *Apis mellifera*. The earliest bee glyphs and most of the Ptolemaic Period bees are highly stylized (Figure 7.4e–g). Many other insects are depicted in Egyptian reliefs, amulets, and jewelry, such as scarab beetles, grasshoppers, flies, mantids, butterflies, metallic wood-boring beetles, and click beetles (Kritsky 1993). Occasionally, some of the insects can be identified as particular species, but many more, such as the grasshopper carved in the walls of Medinet Habu, the mortuary temple of Ramesses III (Figure 7.7), can only be traced to a common name or family. Ransome (1986) reviewed the issue in her 1937 book, *The Sacred Bee*, and wrote that the glyph's inclusion in the apicultural scene at Newoserre Any's sun temple established that the Egyptians were thinking of a honey bee when they depicted the insect. Moreover, its inclusion on amphorae of honey in the tomb of Rekhmire and as part of the honey-related scenes in the tombs of Pabasa and Ankhhor further supports the idea the glyph was intended to represent a honey bee. Considering that the glyphs were painted or carved by artists and artisans rather than beekeepers, it is likely that the workers who created the glyphs were not intimately familiar with bees, and their representations were based on a stylized

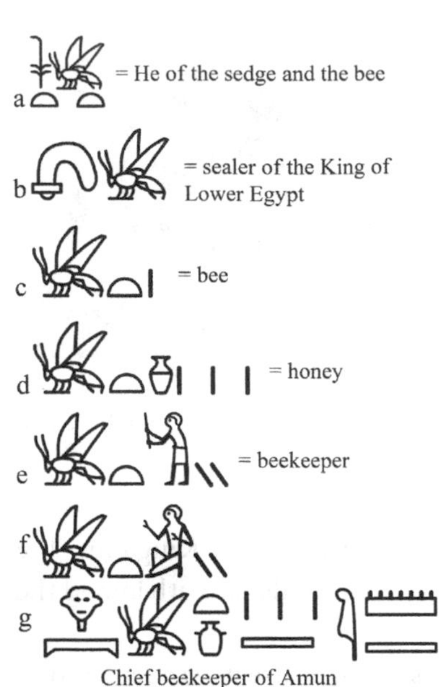

Figure 7.6 Egyptian words and phrases that incorporate the honey bee hieroglyph.

Figure 7.7 The grasshopper hieroglyph from Medinet Habu, the mortuary temple of Ramesses III. Photograph by Gene Kritsky.

convention, not on first-hand observation. If there was any confusion on the part of the early Egyptians, it was probably the same difficulty that most people today have in telling the difference between bees and wasps. Indeed, I am often asked to identify "bees" that are actually wasps, and vice versa.

An Ancient Egyptian Beekeeping Lexicon

It is beyond the scope of this book to examine all the nuances of interpreting hieroglyphs, but knowing what to look for can help verify whether the honey bee hieroglyph is referring to a specific place or to some aspect of honey production. Hieroglyphs can be read either left to right or right to left, as indicated by which way the people and animals in the text are facing: the hieroglyphs are always read toward the faces of people and animals (Strudwick 2010). This may seem strange to us, but it permits hieroglyphs to be inscribed in orderly and symmetrical patterns.

There are hundreds of hieroglyphs in the ancient Egyptian language. The phonetic alphabet consists of a set of 24 consonants, which, like our alphabet, can be combined to produce words. To make matters more interesting, the Egyptians also had biliteral and triliteral hieroglyphs—one symbol that could denote two or three letters, respectively. For example, the loaf glyph in the royal titulary is part of the alphabet and represents the consonant sound *t*. The sedge plant in the titulary is a biliteral sign and is the sound *su*. The hieroglyph of the scarab beetle is a triliteral *ḫpr*. These hieroglyphs can be combined to produce sound combinations or words, but the language does not stop there. The Egyptians also used ideograms (hieroglyphs that convey words or concepts). For example, the hieroglyph for the word "star" is indeed a five-pointed star (Zauzich 1992).

The most common usage of the honey bee hieroglyph and the easiest to interpret is its use in the royal titulary. Visitors to the British Museum who carefully inspect the Rosetta Stone will find the royal titulary six rows down from the top and in the center of the row adjacent to the cartouche for the pharaoh Ptolemy V. The Rosetta Stone was found in 1799 in the village of Rashid, called Rosetta by Europeans. It was part of a wall that French soldiers were going to destroy in order to build a fort. One of the officers recognized that there were three different languages on the stone and surmised that they were likely all describing the same text: hieroglyphs appeared at the top, demotic in the center, and Greek at the bottom. (Demotic is a highly modified form of hieratic script, which is essentially the cursive form of hieroglyphs.) The honey bee hieroglyph compared to the hieratic symbol for the honey bee is a tidy example of how hieroglyphs could be quickly and easily written on papyri (Figure 7.8). The comparison of the Greek with the hieroglyphs helped to make the translation of the ancient Egyptian language possible (Andrews 1981). The four symbols of the royal titulary are written in Greek in the lower section, and the Greek phrase that corresponds to the titulary is translated as "King of Both the Upper and the Lower Lands."

Figure 7.8 The honey bee hieroglyph (left) with its hieratic counterpart (right). Other examples of hieratic honey bees are shown in Figures 8.2 and 8.3.

Being armed with the knowledge that the royal titulary is associated with a royal cartouche makes tracking down this phrase in various tombs and temples quite easy. In fact, at some temples like the Ramesseum or Medinet Habu, it may seem that it is everywhere.

The honey bee hieroglyph in the titulary (Figure 7.6a) refers to Lower Egypt of the Delta region, but it is also used in various titles that might also be pertinent to the Delta. Visitors to Pabasa's or Ankhhor's tomb should be on the lookout for the bee next to the hieroglyph for a cylinder seal, S 19 (Figure 7.6b). In this case, the two symbols refer to the title "Sealer of the King of Lower Egypt." The sealer could also be thought of as a treasurer.

Naturally, the honey bee hieroglyph is also used when referring to bees and bee products. The most obvious example is the word "bee," which is formed by three glyphs (Figure 7.6 c): a honey bee, a loaf, and a single stroke or a wooden dowel. When the dowel appears after an ideogram, it indicates that the ideogram signifies the actual object portrayed in the hieroglyph. In this case, the three glyphs mean "bee." In the Egyptian language, the word for bee is pronounced *bỉt*; the loaf is the sound *t*. Egyptians also referred to the honey bee as the "fly of honey" (Ransome 1986).

The word for honey is built on the word for bee. In addition to the honey bee and a loaf, the hieroglyphic word includes some type of vessel and three strokes (Z 2) rather than one (Figure 7.6d). The multiple strokes make the ideogram plural (Gardiner 1988). The word "honey" was found on the last section of the beekeeping relief at Newoserre Any's sun temple (Figure 2.4). It can also be found in the first delineated column of hieroglyphs on the left above the seated Rekhmire (Figure 4.7); note that the vessel used in the word "honey" is a diamond-shaped vessel similar to those in the beekeeping scene (Figure 4.6). The word "honey" with a different vessel can be found in the Twentieth Dynasty stele of Hori (Figure 7.9). In this case, the vessel used in the word "honey" also indicates the size of the container. The phrase is translated as "honey, *mnt*-jar" (Janssen 1963).

The hieroglyphs used for the word "beekeeper" is similar to the one for "honey," with the substitution of a male figure (A 24 or A 3) for the vessel and two diagonal strokes (Z 4) instead of the three vertical ones (Ransome 1986) (Figure 7.6e and f) (Mehdawy and Hussein 2010).

Funerary cones for officials in charge of beekeepers also incorporate the honey bee hieroglyph. The cone for Neferkhaut, whose tomb is yet to be discovered, has four rows of hieroglyphs that are interpreted as "Revered One before Osiris, Overseer of the Fields of Amun, Chief Beekeeper of Amun, Wab-Priest, Neferkhaut Justified." The third line (Figure 7.6 g) includes the hieroglyphs D 2, Q 5, L 2, X 1, W 23, Z 2, N 1, H 6, and Y 5 (Zenihiro 2013). The last three glyphs are the god's name, "Amun."

Figure 7.9 The Twentieth Dynasty stele of Hori with the word "honey" highlighted in the box. The phrase as written translates to "honey, *mnt*-jar." Photograph by Gene Kritsky. © Trustees of the British Museum.

How the Honey Bee Hieroglyph Was Carved

Ancient Egyptian art changed little during its 3,500-year history. Indeed, the stability of Egyptian art reflected the stability of its society. Except for the Amarna Period, when Akhenaten challenged the religious status quo, Egyptian art exhibited relatively minor changes from the Old Kingdom to the end of the Late Period (Müller 2001a). The consistency of hieroglyphs as well as depictions of gods, humans, and animals was aided by a system of lines that were painted onto walls to guide the proportions of figures and the placement of hieroglyphs. Horizontal or vertical lines were used from the Fifth Dynasty on, and a grid system, introduced during the Eleventh Dynasty, helped to further constrain proportions and formed the underlying structure of Egyptian art in general (Robins 2001). These grid lines were produced in a fashion similar to modern chalk lines used to line up wallpaper or to lay roof shingles: the Egyptian draftsmen would coat a string with pigment (usually red), stretch it taut across the wall, and snap it to produce a straight, clean line. Placed at carefully measured intervals, these lines and grids would serve as guides for painting hieroglyphs and the outlines of figures. (These grid lines can still be seen in the tomb of Ramose (Figure 7.10).)

Royal cartouches were normally produced within a two-by-four grid, and the *nesu-bity* royal titulary was carved to fit a two-by-two adjacent grid (Figure 7.11) (Robins 1994). Ideally, the horizontal centerline of the titulary's four squares was tangential to the top of the bee's head and thorax while the vertical centerline bisected the thorax and the forewing, which ended at the outer edge of the grid. The hind wing cut diagonally across an upper square, ending at or near the corner. The abdomen extended into the lower square below the hind wing and ended near the midpoint of the outer edge of the lower square. The front and middle legs extended forward into the square opposite the abdomen.

Figure 7.10 Grid lines on the walls of the tomb of Ramose were employed to create tomb paintings or reliefs of consistent sizes and proportions (Robins 1986). Photograph by Gene Kritsky.

Figure 7.11 The cartouche of Ramesses III and the royal tituarly with a superimposed grid. Photograph and modifications by Gene Kritsky.

The orientation of the wings and the abdomen within their grid could vary within a single temple. For example, the two wings might both touch the top of the grid, or the wings could be farther apart, with the hind wing touching the side of the upper grid. The abdomen might be carved so that it lined up above the legs, or it could be carved extending below the legs. This variation in the depiction of the honey bee may have resulted from several draftsmen working in different parts of the temple, or from attempts to make the hieroglyph fit into a smaller space.

CHAPTER 8

The Administration and Economics of Egyptian Beekeeping

When most people think of the great accomplishments of ancient Egypt, they envision massive building projects such as the Pyramids of Giza and the Great Sphinx. These spectacular monuments have endured for millennia, but as impressive as they are, they would not have been possible had the Egyptians not discovered how to organize people through an administrative structure.

The pyramids were built by a surprisingly small number of workers. The working group, or team, consisted of just 20 workers with one leader. Ten such teams formed a division of 200 workers, and five divisions formed a gang of 1,000 workers. Two gangs formed a crew, the largest group in the construction organization. Although we do not know for certain how many workers were involved in building the pyramids, estimates place the number between 20,000 and 25,000. Incredibly, it took these workers only 20 years to complete the Great Pyramid of Giza (Wilkinson 2010), and the true accomplishment and one of ancient Egypt's greatest successes is the civil organization required to complete this mammoth task. However, Egyptian overseers did not apply this sophisticated organization solely to construction projects. They also organized people to keep bees.

Exactly how beekeepers were organized in ancient Egypt is not known, but we have ample evidence that they were organized. I have already mentioned Nykara, who was "Overseer of All Beekeepers," and the Middle Kingdom scarab seal with the title "Chief Beekeeper" inscribed on its base. Several other titles suggest a hierarchal organization, and a survey of these apicultural titles throughout Egyptian history illustrates the complexity of that organization. There were beekeepers, chief beekeepers, overseers of the beekeepers, and overseers of the beekeepers of the entire land. There were sealers of the honey, collectors of honey, and beekeepers associated with temples. All of the supervisors ultimately answered to the vizier, who reported to the pharaoh (Brewer et al. 1994).

Beekeepers were also organized by locality. Some beekeepers worked in and around settlements, but there were also specialists who apparently collected honey from wild bees in marginal areas (Shaw and Nicholson 1995). Nykara had a number of duties, many of which concerned overseeing the marginal areas away from larger settlements (García 2010). These remote areas apparently required some additional help to protect the honey harvest. During the Twentieth Dynasty, archers were appointed to protect "collectors of honey" (Breasted 1906b: 175).

Other beekeepers directly worked with temples and supplied them with honey for ritual needs. One individual was the actual beekeeper, and another was more concerned with gathering and possibly processing the honey. The Egyptian collection in the Neues Museum in Berlin includes the unfinished stele of Smentu, who was "Head of Beekeepers of His Lord, before Min and Isis." The round-topped stele of Khons at the Kestner Museum in Hanover, Germany, includes his title as "Beekeeper of Amun Great of Victory." During the reign of Ramesses IV in the Twentieth Dynasty, Ouser-mare-nakhty held the title of "Purveyor of Honey to Min, Who is in the Laboratory at Coptos." Min was the god of physical reproduction and Coptos was the center of the Min cult (Brewer et al. 1994). A New Kingdom mud-brick tomb discovered in 2002 at Saqqara further illustrates temple beekeeping; the tomb belonged to Ibi, whose title was "Overseer of Honey Production in the House of Amun" (Eisenberg 2002).

Additional evidence of beekeepers associated with temples comes from the study of funerary cones—conical pottery objects that were incorporated into the façade of a tomb. During the New Kingdom, the base of a cone was inscribed with the name of the tomb's owner, with the owner's titles, and often with protective spells. A few cones have been found that identify a tomb belonging to an official associated with beekeeping. For example, two such cones were found at the tomb of Mery (TT95), an Eighteenth Dynasty official. One cone (#390) reads, "Beekeeper, Overseer of the Prophets in Upper and Lower Egypt, First Prophet of Amun, Mery, Steward of Amun, Overseer of the Double Granary of Amun, Mery, Overseer of the Double Treasury (House of Silver) and Overseer of the Double House of Gold of Amun, Mery, Overseer of the Cattle of Amun, Mery." A second cone (#400) from Mery's tomb reads, "Beekeeper, First Prophet of Amun, Overseer of the Prophets in Upper and Lower Egypt, Mery, Overseer of the Fields of Amun, Overseer of the Double Granary of Amun, Mery, Sealer of Everything in the House of the King, Mery, Overseer of the Cattle of Amun, Mery. Another funerary cone (#455) from the yet-to-be-discovered tomb of Minmose states, "Osiris, Beekeeper of Amun, Bearer of Incense, Minmose Justified, Possessor of Honor [Before] the Great God (?)." Cone #83 from Imiamun's tomb has the inscription, "Revered One by the Great God, Overseer of the Beekeepers of the King of Lower Egypt, Imiamun, Justified before Osiris" (Zenihiro 2013).

There is also evidence as to how the apicultural chain of command may have worked. In the Ashmolean Museum at Oxford University is a Nineteenth Dynasty letter from a scribe complaining that two beekeepers had not met their honey quota. The letter noted that one of the beekeepers was continuing to work even though he had been dismissed. The scribe argued that his overseer should take stronger action in this case (Crane 1999).

Several papyri from the Ptolemaic Era indicate that the organization of beekeepers continued while Egypt was being governed by outsiders. In 1914–1915, papyri belonging to the Greek official Zenon, who lived in the Faiyum around 250 BCE, were discovered. Among those papyri was a written request from beekeepers:

> To Zenon greeting from the beekeepers of the Arsinoite nome [a political district]. You wrote about the donkeys, that they were to come to Philadelphia and work ten days. But it is now eighteen days that they have been working and the hives have been kept in the fields, and it is time to bring them home and we have no donkeys to carry them back. Now it is no small impost that we pay the king. Unless the donkeys are sent at once, the result will be that the hives will be ruined and the impost lost. Already the peasants are warning us, saying: "We are going to release the water and burn the brushwood, so unless you remove [the hives] you will lose them." We beg you then, if it pleases you, to send us our donkeys, in order that we may remove them. And after removing them we will come back with the donkeys when you need them. May you prosper! (Ransome 1986: 27)

Another papyrus from the Zenon archive mentions an individual named Samos, who owned 5,000 hives in several nomes. Still another letter from two Greek beekeepers reads, "We owned already under the present King's father 1,000 hives, which were leased out" (Bingen 2007). Finally, a third papyrus, this one to Herostratus from two beekeepers in Oxyrhynchus, reads, "We possess 487 beehives, of which we had 87 in the village of Toka" (Crane 1999: 167).

Beekeepers continued to have some form of organization in the Roman Era, as documented in the Oxyrhynchus papyri. The Oxyrhynchus papyri were discovered at the end of the 19th century in an ancient trash dump, and they were written between the first and the sixth centuries CE. Oxyrhynchus papyus 85 was written in 338 CE and involves the declaration of a variety of guilds, including coppersmiths, brewers and beer sellers, and bakers. The declaration described a formula that would permit an assessment or a valuation of the goods in stock at the end of the month. The papyrus ends with a declaration for the beekeepers' guild, but it is incomplete. However, it does provide further evidence that beekeeping remained a profession and that there was some kind of organization to provide a common administration (Grenfell and Hunt 1898). All of the communications support the idea that there was a central beekeeping authority, who could dismiss beekeepers, provide resources, and receive reports on the status of apicultural operations.

The honey that was produced by various beekeepers was graded based on color and purity. Kuény (1950) suggests the grades were *stf* for light-colored honey, *dsrt* for red or desert honey, and *pw-ǵ mh-tt,* a classification that is not well understood (Brewer et al. 1994, Crane 1999). There is a stamp at the New Museum of Ancient Egyptian Agriculture in Dokki that Crane (1999) suggested may have been used as an official stamp for grading. This stamp (Figure 8.1) shows four bees arranged in pairs facing each other, mirrored across a horizontal axis. Between the four bees in the middle of the stamp is a man with his arms held to his sides.

The administrative duties of some officials involved in beekeeping included the accounting of honey tributes and payments (Figure 8.2), and there are numerous examples of labeled jars and seals to suggest that there was some kind of inspection or labeling. A famous example from the tomb of Tutankhamun is part of a two-handled amphora with the word "honey" clearly written in hieratic on its surface. The Petrie Museum in London possesses several examples of seals and potsherds that reference honey. UC59538 is a plaster seal dating from the time of Seti I, bearing part of the phrase "honey of the Menmaatre

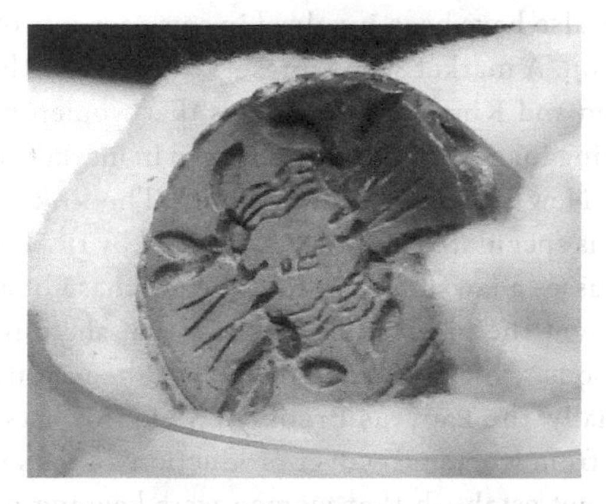

Figure 8.1 A stamp at the New Museum of Ancient Egyptian Agriculture in Dokki; it has been suggested that it was used to grade honey. Photograph by Beth Cortright and used with permission.

Figure 8.2 A schematic showing the honey tributes recorded in the tomb of Rekhmire. From Newberry (1900).

temple on the west of Thebes." The potsherd, UC32262, is a piece of an Eighteenth Dynasty Canaanite amphora with "honey" written on it (Figure 8.3). The potsherd, UC32259, also from the Eighteenth Dynasty, refers to a delivery of honey, and UC46297 is a potsherd from a storage jar dating from the Amarna period of Pharaoh Akhenaten with a honey bee hieratic glyph, indicating that it held honey (Figure 8.4).

All of the tomb reliefs and paintings show only men working the hives, extracting, and sealing honey. However, there is some evidence to suggest that women may also have been involved in some aspects of beekeeping or honey distribution. A market scene, preserved in the Fifth Dynasty tomb of Niankhkhnum and Khnumhotep, shows that women were engaged in selling or bartering for goods. Women engaged in market selling were also recorded in the New Kingdom tomb of Ipuy. The link between women and beekeeping is speculative, but it is suggested by the Cairo papyrus 65 739, which discusses a lawsuit regarding the purchase of a slave girl. The woman who bought the girl wanted to show that she purchased the girl with her own goods, which included cloth that she produced and a jar of honey. Additionally, the Papyrus Brooklyn 35.1453A lists the delivery of yarn and honey from women to cover production deficits from a previous year. This does not establish that women were keeping bees, but it does support the idea that women may have functioned as dealers in cloth and honey (Eyre 1998).

Figure 8.3 An Eighteenth Dynasty pottery potsherd from an amphora labeled with the word "honey" UC32262. Photograph by Gene Kritsky and published courtesy of the Petrie Museum of Egyptian Archaeology, UCL.

Figure 8.4 A potsherd from an amphora labeled with the word "honey" from the Amarna period UC46297. Photograph by Gene Kritsky and published courtesy of the Petrie Museum of Egyptian Archaeology, UCL.

The Value of Ancient Egyptian Honey

Ancient Egyptian economics appear alien to those of us who live in a currency-based economy. Before the Ptolemaic and Roman periods, the Egyptians did not have a single currency or coinage; there was no "money," as we understand it. Rather, their economy was based on distribution of goods based on the commodities' value. The yardstick applied for valuation of goods was the *deben*. One *deben* was a specific weight of copper, silver, or gold. In most cases, the comparative value for honey was the value of 1 *deben* (91 grams) of copper (Bleiberg 2001). The *deben* was not based on the intrinsic value of the copper, but rather on what could be made with the metal. Thus, 20 *deben* of copper could make a copper vessel, and that was its value. One *deben* of copper was valued at 1 *kite* of silver, which was one-tenth of a *deben* of silver (Janssen 1975). Another unit of value that was used during the Nineteenth and the early Twentieth Dynasties was called the *senyu*. This was 7.6 grams of silver, which was one-twelfth the weight of a *deben*; it was worth 5 *deben* of copper (Bleiberg 2001). These values did not remain constant throughout Egyptian history. For example, during the reign of Ramesses II in the Nineteenth Dynasty, 1 *deben* of silver was worth the equivalent of 100 *deben* of copper. However, during the reign of Ramesses IX about 180 years later, 1 *deben* of silver was worth 60 *deben* of copper.

People did not walk around carrying these quantities of metal. Instead, workers would be compensated by receiving goods of comparable value. If a worker was to be paid the equivalent of 2 *deben* of copper, he or she would likely receive a sack of grain that would have the value of 2 *deben*, not the metal itself (Bleiberg 2001). Accordingly, the value of the *deben* is helpful in establishing the barter value of honey.

Honey was measured by volume, but the Egyptians did not have a Bureau of Standards that we can consult for approximations of their units of volumetric measure. Honey was usually transported in *mnt* or *mḏḳt* jars. However, the *hin* was the unit of measure for smaller quantities (Janssen 1975). Three Eighteenth Dynasty alabaster vessels in the Museum of Leyden are labeled 25, 12, and 7¼ *hin*. These vessels were filled with water to determine the volume in liters, and the average of the three vessels indicated that 1 *hin* contained about 0.46 liters. Another vessel from the Nineteenth Dynasty that was marked 40 *hin* also showed that 1 *hin* was equivalent to 0.46 liters. Measurements of similarly marked vessels in the Cairo Museum, the Turin Museum, and the British Museum indicated that the *hin* was 0.425 liters, 0.412 liters, and 0.534 liters, respectively. Based on these totals, it is generally accepted that a *hin* was approximately 0.48 liters (Sobhy 1924).

Once the volume of the *hin* is known, it is possible to estimate the quantities held by *mnt* or *md̲k̩t* jars. One *mnt* jar contained approximately 40 *hin* and the *md̲k̩t* contained 50 *hin*. During the time of Ramesses II, one *mnt* jar of honey was worth 5 *kite* of silver or 50 *deben* of copper. Thus, 1 *hin* of honey was valued at 1.2 *deben* of copper.

The value of honey fluctuated in ancient Egypt, as it does today. During the Nineteenth Dynasty when Ramesses II was ruling Egypt, one *hin* of honey was valued at 1.2 *deben*. In the following dynasty, the value of honey dropped. A papyrus from the seventh year of the reign of Ramesses IX, a century after Ramesses II, recorded that a *md̲k̩t* jar of honey was valued at 38 *deben* (38 *deben*/50 *hin*), or 0.76 *deben* per *hin*. This represented a decline of nearly 37% between the dynasties (Janssen 1975).

An illustration of Egyptian honey economics can be found in an ancient Egyptian letter that was found in 1907 by the French Egyptologist Jean Clédat. The letter was pieced together in 1975 from fragments of papyri and subsequently translated. It describes an apparent theft of honey and was sent to the governor of Elephantine, Montuherkhopshef, from the scribe Khay. It reads:

> Khay of the domain of Harakhty gre[ets the governor] of Elephantine. [Montuherkhopshef, in life, prosperity], health, in the favor of Amun-Re, king of the gods. I say to Amun-Re, to Har[akhty] when he rises and sets, to Harakhty and his ennead, that they make you well, that they make you live, that they make you be in the favor of Harakhty, your lord who watches over you.
>
> I opened the jars of honey that you brought for the god. I removed ten *hin* of honey for the divine offering. I noted that they were all filled with bars of unguent. I closed them again. I had them taken south. If it is someone else who gave them to you, make him answer for it, and see if you find a good honey, and have it brought to me. And if there is none, let a *mnt*-jar of incense be brought to me by the priest Netermose. (Vernus 2003: 101, Posener-Kriéger 1978: 85)

Incense is normally transported in *mnt*-jars, which would have a value of between 7 and 7½ *deben* of copper. The 10 *hin* of honey at the time the letter was written would have been valued at between 7 and 8 *deben*. This was typical of how exchange was carried out: there was a known value of a particular commodity, and if the desired form of payment was not available, a substitute of equivalent value would be requested or offered (Posener-Kriéger 1978).

Honey was expensive in ancient Egypt and therefore was not widely enjoyed by the general population. However, there is evidence of individuals in the higher classes having some access to honey (Brewer and Teeter 2007). In the Eighteenth Dynasty tomb of Ineni, an architect and government official, is the inscription, "I was supplied from the table of the king with bread of oblations for the king, beer likewise, meat, fat-meat, vegetables, various fruit, honey, cakes, wine, oil. My necessities were apportioned in life and health, as his majesty himself said, for love of me" (Breasted 1906: 47–48).

During the Nineteenth Dynasty, the Silsilah quarry stele recorded the king's messenger and standard-bearer's rations as "good bread, ox-flesh, wine, sweet oil, [olive] oil, fat, honey, figs, fish, and vegetables every day" (Breasted 1906a: 90). Honey does not seem to have been provided to the quarry workers (Brewer et al. 1994).

CHAPTER 9

Bees and Food

The ancient Egyptians excelled at horticulture. Tomb reliefs and excavations document an abundance of food crops, including barley, wheat, dates, figs, grapes, heglig (desert dates), persea (Egyptian avocado, *Mimusops* spp.), pomegranates, olives, apples, peaches, pears, and cherries. Reliefs and paintings show that they planted formal gardens, intercropping figs, olives, and other flowering plants and watering them with *shadufs*, an irrigation system of a container placed at one end of a pole with a counterweight at the other, permitting the gardener to easily dip the container into the Nile and lift it out with little effort.

Planting the variety of food crops raised in Egypt required a vast knowledge of propagation techniques. Date palms were started by rooting basal offshoots, whereas the sycamore fig was started by rooting cut twigs. Other plants were grafted, started from bulbs, or planted as seeds (Murray 2000). Further evidence of the ancient Egyptians' horticultural sophistication is demonstrated by their cultivation of the sycamore fig. The wasp required for the pollination of the imported plant was not present in Egypt, which required the growers to induce ripening by scarification—scoring the rind of the immature fruit. This wound caused the fruit to release ethylene, inducing the immature fruit to quickly grow and ripen into a parthenocarpic (unfertilized) fruit (Janick 2002). Did this knowledge also extend to an understanding of the role that bees played in food production? Did ancient Egyptian gardeners understand the concept of pollination?

The Salt 825 papyrus, written approximately 2,300 years ago, indicates the Egyptians knew that there was a link between bees and plants. The papyrus reads that the bee "busied himself with the flowers of every plant" (Leek 1975: 148). Other translations of the Salt papyrus seem to confirm the association of the honey bee with flowers. Manassa's (2008: 115) more recent translation reads, "As soon as all the bees had been fashioned, its work in the flowers of all the fields came into existence." Crane (1999: 602) translated Derchain's French interpretation as "its task was [to work on] the flowers of every plant."

Moreover, the next sentence of the Salt papyrus shows that the Egyptians knew that the bees' work on the flowers was necessary for honey production. The various translations read, "and so wax was made and also honey" (Leek 1975: 148); "it means that beeswax came into being; it means that honey came into being from its liquid" (Manassa 2008: 115); and "that is how wax came to be, and how honey came to be" (Crane 1999: 602).

The honey bee–plant association may be central to the interpretation of the upper right register of the tomb of Ankhhor (Figure 5.16). What remains of the relief shows a man holding his hand in a gesture summoning the bees on the other side of the tree. This relief may be the earliest recognition that honey bees worked plants in order to produce honey, predating the Salt papyrus by approximately 350 years.

There is no direct evidence to suggest that the Egyptians understood that bees were needed for pollination. Crane (1999) indicated that the pollen balls brought back to the hives by the bees were thought to be wax, and this would be consistent with the details of the Salt papyrus suggesting that bees working the flowers were collecting wax. Moreover, even though the Egyptians understood that sycamore figs would not ripen without scarification, they clearly did not know that this was due to the absence of the fruit's pollinator, the fig wasp. The fig wasp is tiny, the pollination cycle is complex and cryptic, taking place inside the fruit, and it did not occur in Egypt.

Zander (1941) described honeycomb that was found in a Nineteenth Dynasty tomb at Deir el-Medina, also called the Valley of the Workers. The comb pattern indicated that it was made by *Apis mellifera*, and Zander went on to suggest that it belonged to the subspecies *unicolor fasciata*. The contents of the honeycomb were dissolved in water and found to contain mostly the pollen of *Mimusops schimperi* and *Balanites aegyptiaca* (Egyptian avocado and desert dates, respectively). He also detected the pollen of *Graminae, Caryophullaceae, Trifolium, Melilotus, Vicia, Cruciferae, Rosaceae, Thymelaceae,* and *Polygalaceae,* which indicate that the plants of Egypt have changed considerably since the time of the pharaohs.

The presence of pollen in Egyptian containers without honeycomb has provided the best evidence that they once contained honey. There have been several reports that still-edible honey was found in jars and vases from ancient tombs, but these claims have not been corroborated by chemical analysis. Samples of dark, thick substances dating from the Middle and New Kingdoms have been verified as honey from the pollen grains they contained (Crane 1999).

Honey in Food

Honey was the major sweetener in the ancient world, as sugarcane had not yet been introduced from Southeast Asia. The Egyptians consumed honey directly, but it was also used in making honey cakes. In the Eighteenth Dynasty tomb of Rekhmire, just to the right of the beekeeping scene is a scene that shows the manufacture of *shat*-cakes or honey cakes. Amphorae labeled as honey are included in the scene, providing the evidence of honey included as an ingredient. These cakes were shaped into long triangles or wedges and appear to have been covered with sesame seeds (Murray 2004) (Figure 9.1). Wilson (1988a) wrote that the cakes were made from a puree of dates and honey, and that these would have been a luxury food. According to other scenes from his tomb, Rekhmire was very fond of a cake made from tiger nuts (the small tubers of a sedge, *Cyperus esculentus*); the cakes consisted of flour ground from the tubers, mixed with honey, and baked (Brier and Hobbs 2009).

A complete dinner was preserved and placed in a Second Dynasty tomb from Saqqara. The menu items included bread, barley porridge, fish, pigeon soup, quail, kidneys, beef, an assortment of fruits, cheese, wine, and pies made with honey. Honey pies are still popular in Egypt (Mehdawy and Hussein 2010).

Figure 9.1 A schematic from the tomb of Rekhmire showing the making of honey cakes. From Newberry (1900).

Honey was also added to breads during the Greco-Roman Period. A popular bread in Alexandria was *pankarpian,* which was made from sieved grains to which honey was added. The mixture was then rolled into balls and wrapped in papyrus for cooking (Mehdawy and Hussein 2010).

Honey was also used to sweeten wine and beer (Darby et al., Ikram 1994). One illustration of this is a story told by a prisoner, recorded on a papyrus in the British Museum:

> I heard, when I was imprisoned with the warehouseman Iufenamen, that silver had been given to Nespare in exchange for beer.
>
> The thief Amenkhau gave one *deben* and five *kite* of silver to the acolyte Paenimentet in exchange for a measure of honey; the incense preparer Paenimentet said that he gave him (= Amenkhau) another measure of honey, and Amenkhau, the thief, gave him in exchange one *deben* and five *kite* of silver. Total: three *deben* of silver.
>
> The thief, the young slave Amenkhau son of Mutemheb, gave five *kite* of silver to the scribe of the overseer of the domain of Amun Aasheftemwese in exchange for a [unknown quantity] of wine; we took it to the house of the overseer of peasants: we added two *hin* of honey to it and we drank it. (Vernus 2003: 36)

It is surprising, given the abundance of honey, that the Egyptians did not develop mead (Ikram 1994). McGovern (2009) thinks that this may be due to the presence of beer and wine early in Egypt's history.

It has been suggested that honey may also have been used as a meat preservative. The evidence for this is from an Eighteenth Dynasty tomb belonging to Ahmose Meriamun. Ducks, geese, and pigeons from the tomb may have been thinly wrapped in linen that had been soaked in some form of a resinous solution, which may have included honey. Despite this find, most authorities do not think that meats were normally preserved in honey, as it was an expensive commodity (Ikram 2000).

CHAPTER 10

Honey and Healing

Honey was an important sweetener for food and wines, and a commodity used to pay tribute and taxes. However, its use in ancient Egyptian medicine was one of its most significant applications.

Medicine in ancient Egypt was highly regarded in the region. Some medical texts are confirmed to date back to the Twelfth Dynasty, and there are other medical papyri that may date to the Old Kingdom. The Egyptian physicians' knowledge of plants and their uses in treating all sorts of diseases, especially by the time of the Persian conquest of Egypt, was considered among the best in the world. Even Hippocrates and Galen acknowledged that part of their medical knowledge came from Egypt (Hornsey 2012).

Wealthy Egyptians had access to a diverse diet of grains, fruit, leafy vegetables, fish, poultry, cattle, milk, wine and beer, and, of course, honey. The lower classes primarily consumed grain and beer, which was payment for wages. Life expectancy ranged between 30 and 36 years for most people, and a few Egyptians, such as Pepy II and Ramesses II, lived into their 80s (Ritner 2001a).

The physicians who cared for the Egyptian population sought the assistance of the various gods involved with protection. Thoth, the ibis-headed god, was the god of healing; the lion-headed goddess Sekhmet could bring pestilence and illness (De Jong 2001), but her alter ego, Bast, was associated with nurturing and protection from harm. Given this link between medicine and the gods, medical training may have been centered in the temple. Like the administrative structure of the beekeepers, there appears to have been a similar organization of physicians, as documented by the several references to medical titles that included physicians, overseers of the physicians, chief physicians, palace physicians, and even inspectors of physicians. Moreover, Egyptian physicians could specialize in dentistry, ophthalmology, pharmacy, internal disease, urology, and gynecology. Doctors were also trained in trauma medicine and used splints, sutures, and even trephination (Ritner 2001a). At Kom Ombo, there is a relief of medical instruments (Figure 10.1) including hooks, saws, forceps, scalpels, and a sink (Nunn 1996). In keeping with the association between

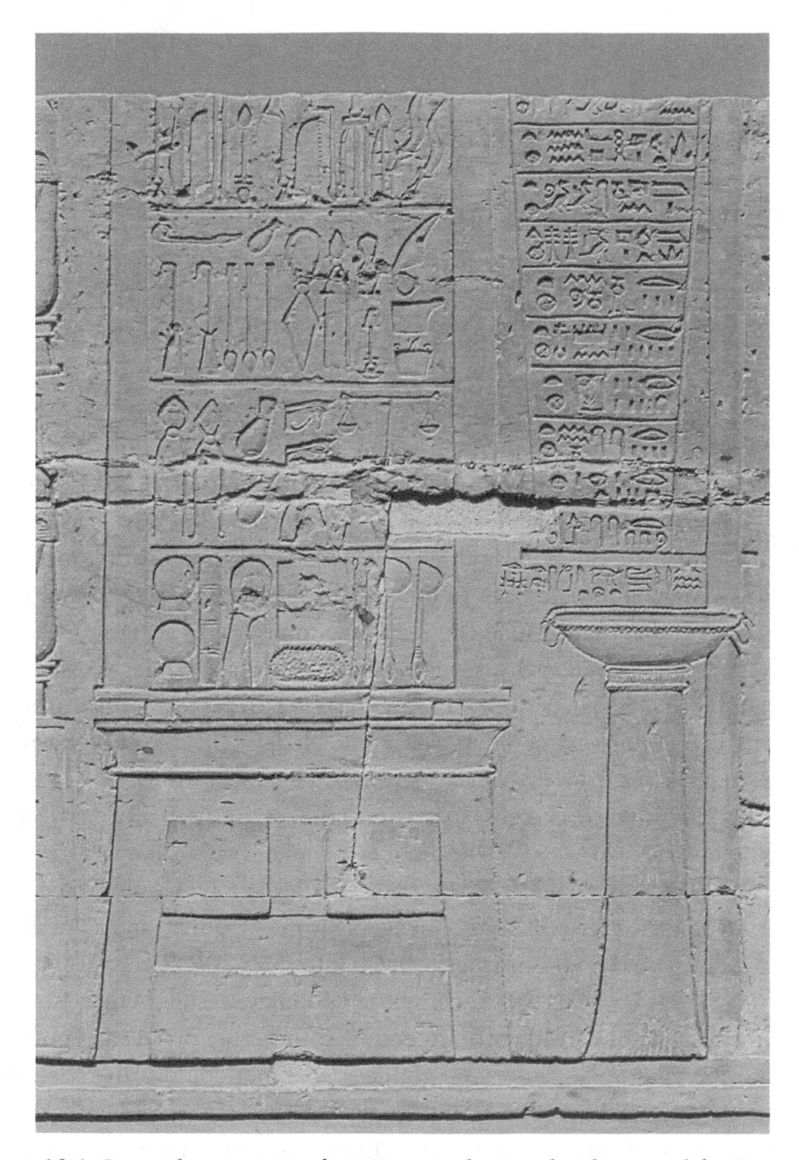

Figure 10.1 Surgical instruments from Kom Ombo temple. Photograph by Gene Kritsky.

physicians and religion was the use of magic to treat different ailments. For example, physicians might prescribe a "comforting spell" or recommend the use of an amulet as part of the therapy (Ritner 2001).

The apicultural interest in ancient Egyptian medicine comes primarily from the papyri that have survived (Figure 10.2). During the second century CE, Clement of Alexandria enumerated 42 Egyptian books of human knowledge. Of these books, six were dedicated to medical knowledge. Clement

Figure 10.2 A papyrus with a medical prescription that includes honey. Photograph by Gene Kritsky. The Louvre, Paris.

listed contents of the six books: anatomy, diseases, physicians' tools, remedies, diseases of the eye, and gynecology. These books do not survive intact, but there are several papyri that deal with these six medical topics. The oldest is the Kahun papyrus, which dates back to the Middle Kingdom and pertains to gynecology. The Smith papyrus covers trauma and surgery, and the Ebers papyrus covers a wide range of general medicine. Honey plays a major role in many of the treatments discussed in these medical papyri. Over 900 treatments are mentioned in the various medical papyri, and of those, about 500 use honey as an ingredient for the treatment of respiratory problems, digestive problems, and parasites (Sagrillo 2001). In some cases, honey was used primarily to make the medicine more palatable or to act as a binder to promote the action of other ingredients.

Besides honey, there are numerous references to the use of beeswax in Egyptian medicine. Just over 4% of the recipes in the Ebers papyrus and about 13% of the treatments mentioned in the Hearst papyrus include beeswax. In these cases, the wax is used as a binder for the ingredients, as an adhesive for bandages, or as the active ingredient in the treatment of gut disorders (Estes 1993).

A survey of remedies, mostly from the Ebers papyrus, that include honey and wax provides insight into Egyptian medicine and the role that bee products played in various treatments. The Ebers papyrus, which measures 68 feet long and 12 inches wide, was purchased in 1872 by George Ebers following a sales notice in an antiquities catalog. The papyrus was originally purchased by Edwin Smith in 1862 when he was in Luxor. Allegedly, it was found rolled up between the legs of a mummy in a tomb in the el-Asasif necropolis, the cemetery where Pabasa and Ankhhor were entombed. It was thought that the mummy must have been a physician, but details of where it was found have long since been lost. After Ebers purchased the scroll, he returned to Germany, where he deposited it in the University of Leipzig library (Nunn 1996).

The Ebers papyrus records 877 prescriptions for ointments, inhalations, gargles, draughts, pills, fumigants, suppositories, and enemata. The ingredients encompassed all kinds of animal, mineral, and vegetable matter, including honey, honeycomb, and wax. The Ebers papyrus includes several treatments for constipation, but one listed as a remedy to "open the belly" described a mixture of milk, notched sycamore figs, and honey that was boiled and then strained. The patient was to drink the concoction for four days. Apparently, some of the constipation treatments were so effective that they would result in a prolapse of the anus, which would be uncomfortable. To soothe this condition, the patient would apply a mixture of salt, oil, and honey to the anus for four days (Nunn 1996).

To treat excessive urination, honey would be used, along with groats of wheat, gum, grapes, and ochre. To reduce burning or pain during urination, the patient would drink a mixture of moringa oil, honey, salt, and sweet beer.

The Kahun papyrus described contraceptive pessaries (intravaginal devices) made from honey, natron, sour milk, and crocodile dung. Occasionally, acacia gum would be substituted for the crocodile feces. Another contraceptive treatment described in Ebers papyrus 783 uses ground acacia, carob, and dates mixed with honey as a means to keep a woman from conceiving for up to three years. It is now known that acacia gum and possibly the acidic sour milk could act as effective spermicides. (The efficacy of the crocodile dung remains unproven.)

Other gynecological treatments were also based on honey's putative therapeutic properties. To contract the uterus (although it is not clear if this was to hasten childbirth, to expel the placenta, or to return the uterus to its normal size after birth), a vaginal suppository was made of honey, *kheper-wer* plant, water of carob, and milk. Honey was also used in inducing birth (Nunn 1996).

Honey figures prominently in the treatment of open wounds and burns. It was prescribed for the treatment of an open wound to the skull after the patient showed signs of tetanus. Once the victim was made comfortable, and the jaw began to open, the wound was bound with oil and honey (Nunn 1996). A puncture wound would be treated with a topical ointment made from honey, animal fat, and malachite (Estes 1993). Honey was used to stop bleeding and to promote healing following circumcision. It was also considered effective in reducing inflammation, promoting healing, and soothing the pain from burns and from human bites. Further skin care applications of honey included a topical preparation of ground alabaster, natron, salt, and honey, which was believed to improve the appearance of the skin. To combat excessive perspiration and body odor, the Ebers papyrus recommended an application of honey, red natron, and salt.

The Egyptians used honey for several eye treatments. One in particular was used to remove a foreign object from the eye or to treat eye injuries. On the first day of treatment, the patient was to rinse the eye in marsh water. On the second day, the patient should apply honey and black eye-paint for one day, unless this treatment caused the eye to bleed. If it did bleed, the eye should be bandaged for two days. If there was significant discharge from the eye, an application of cooked green eye-paint, incense, and the top of the *heden*-plant would be used. Honey could also be mixed with finely ground jasper or serpentine and applied directly to the eyelids to cool the eyes (Manniche 1989).

Ancient Egyptian dentistry incorporated honey for various treatments. Generally, the Egyptians suffered from fewer dental caries than people living today, most likely reflecting the lack of refined sugar and the expense of honey. Most dental problems probably came from eating stone-ground bread and the ubiquitous sand. The Ebers papyrus describes using honey for treating a loose tooth; the prescription used crushed seeds, ochre, and honey that were mixed into a paste, which was applied to the loose tooth. For a toothache, a number of plant and mineral substances would be mixed into a honey-based paste to be applied to the tooth (Leek 1967).

In addition to the various medical papyri, there is a container in the National Archaeological Museum of Naples that can further attest to the value of honey as a medicine. The vessel is a small cup measuring 6 centimeters tall and 7.7 centimeters in diameter. Written on the side in hieratic is the phrase "cumin, set milk, honey." These ingredients are known from a number of Egyptian texts, including the Berlin papyrus, which reads, "Another for driving out cough: set milk, cumin; it is to be dipped in honey; the patient is to be caused to eat it for four days." The provenance of the cup is unknown, but given its condition, it is thought to have been part of the grave goods for an individual who suffered from a cough in this life and might need help with it in the afterlife (Poole 2001). If he also suffered from congested sinuses, he might turn to a preparation found in the Ebers papyrus: equal parts galena, petrified wood, dry incense, and honey, which was applied to the nose for four days.

The power of honey as a medicine relied as much on its magical significance as on its physical properties. Some of the prescriptions were accompanied by a spell or incantation that was associated with the power of a god. To treat the *hefat*-worm (possibly the roundworm), the patient ingested sedge and *shames*-plant (possibly pellitory, *Anacyclus pyrethrum*) that had been cooked in honey. This treatment was accompanied by an incantation imploring the god to help reduce the fatigue caused by the parasite, and asking the god to help destroy the infection (Nunn 1996). Similarly, for the treatment of cataracts, the brain of a tortoise was combined with honey and administered as a salve to the eyes while a magical spell was recited. If there was significant discharge from the eye, the tortoise-brain and honey mixture would be combined with the powder from a "statue" (probably an amulet of the deity) and the leaves of a castor tree (Brier and Hobbs 2009).

The Properties of Honey

Honey was a perfect medicine for the ancient Egyptians. Its mythical derivation from the eye of Re imbued it, in the Egyptian mind, with creative power. It made bitter or even foul-tasting substances more palatable. It was viscous, making it ideal for pastes and ointments. Its stickiness ensured that it would remain on a wound if applied as a salve. Through their various preparations, the ancient Egyptians seem to have discovered that honey had therapeutic properties, many of which are now being explored through modern science.

Honey is a processed product of bees, which must have seemed a magical gift in ancient Egypt. It is produced from the nectar that bees collect from flowers. Nectar is essentially dilute sugar water, so the bees must reduce the water content of nectar, assisting its evaporation by regurgitating little droplets of nectar on their mouthparts. As this is happening, other bees create airflow in the hive by beating their wings. This reduces the water content from 80% to 90% down to 13% to 18%, thickening the honey. The bees' saliva contains invertase, which converts the sucrose in the nectar into glucose and fructose. The glucose is converted into hydrogen peroxide and gluconic acid by another enzyme, glucose oxidase. The gluconic acid lowers the pH of honey to the point that it interferes with bacterial growth (Berenbaum 2010, Zumla and Lulat 1989).

Ultimately, honey has a very low moisture content and a low pH, both of which confer antibacterial properties. This is to the advantage of the bees, as the honey can be stored for extended periods of time without spoiling or growing moldy, but it also means that honey is an effective aid in treating cuts, wounds, burns, and any other symptoms tied to bacterial growth. These properties were so important to the Egyptians that honey was used to clean their hands and even the containers that were used to prepare ointments (Hornsey 2012).

Although the ancient Egyptians were unaware of the antibacterial properties of honey (or of bacteria, for that matter), they did recognize its effectiveness as a treatment. Few other ingredients were as widely prescribed by ancient Egyptian physicians as this gift from the tears of Re.

CHAPTER 11

Bees, Gods, and Feasts

Although honey bees may have been created from the tears of the sun god Re, he is not the most ancient Egyptian deity to have an association with bees and honey. The deities Neith, Nut, Min, and Hathor also have links to bees and honey, and their mythologies predate Re. Re, as the sun god and creator, evolved in Egyptian theology between the Second and Fifth Dynasties to his familiar depiction as a falcon-headed man crowned with the sun disk (Müller 2001). The Salt 825 papyrus, which tells the story of Re's creation of the honey bee, is even younger, dating to 300 BCE during the time of the Ptolemies.

The goddess Neith (Figure 11.1) can be traced back to the beginning of the First Dynasty, as evidenced by the name Neithhotep, who was the first queen of Egypt. Neith was associated with the Delta region, which was represented by the hieroglyph of a honey bee. Neith was also called "She of the Lower Egyptian Crown" and was occasionally depicted wearing the Red Crown of Lower Egypt (Simon 2001). Neith was believed to be the mother of all life, including gods and humans. During the New Kingdom, she was thought of as the mother of Re (who in turn created honey bees) as well as the mother of the crocodile god, Sobek.

Although Neith was part of the pantheon from the beginning of Egyptian history, she reached her greatest influence during the Twenty-sixth Dynasty. She was revered in Sais, the capital of Egypt at the time, and her temple at Sais was called *per-bit,* which translated to "House of the Bee" (Sagrillo 2001). Unfortunately, the temple has long since been destroyed, sending it and any reliefs that would further explain Neith's link to honey bees into the sands of Egypt.

Neith's importance during the Twenty-sixth Dynasty may have influenced the beekeeping reliefs found in the tombs of Pabasa and Ankhhor. Unlike the relief at Newoserre Any's sun temple and the painted scene in the tomb of Rekhmire, which show beekeeping and honey-processing activities, the two Twenty-sixth Dynasty tombs illustrate more reverential scenes. The man in the lower left registers of both scenes is holding his hands as if in worship, and

Figure 11.1 The goddess Neith. Photograph by Gene Kritsky. The Louvre, Paris.

the man in the lower right register is making an offering to bees. In the upper left register, a man is pouring a liquid from a container that is the same kind of vessel used on the offering stands behind him. These actions are ritualistic in nature and may reflect a theological link among bees, Neith, and Sais.

"The Myth of Sun's Eye," which dates to Ptolemaic and Roman times, expanded the association of Neith with bees to include Neith calling the queen bee using a reed flute. One myth reads, "When they come to write 'honey,' it is an image of Neith with a reed in her hand that they make." Another myth relates to "beekeepers" using a reed flute "because it is a reed which Neith seized at the beginning" (Widmer 1999).

Another deity associated with bees is the goddess Nut (Figure 11.2), who was the sky goddess; like Neith, she dates back to the Old Kingdom. She is mentioned in the coffin texts as the "Mother of the Gods." Nut is often found on the ceilings of tombs and on the inside of coffin and sarcophagus lids. Nut was a creative force; she swallowed the sun each evening, and gave birth to it every morning. Thus, she was associated with the sun god Re (Lesko 2001a).

The Pyramid texts were the first funerary writings that were carved on the walls of pyramids. They include around 800 utterances or spells, and they

Figure 11.2 The goddess Nut. Photograph by Gene Kritsky. The Louvre, Paris.

described the nightly journey of the afterlife. The oldest known pyramid texts are found in the pyramid of the pharaoh Unas. The texts are of interest to us because they record how Nut appeared as a bee. Utterance 431 is translated as "You are the daughter, mighty in her mother, who appeared as a bee; make the King a spirit within yourself, for he has not died." Utterance 444 continues this association: "O Nut, you have appeared as a bee; you have power over the gods, their doubles, their heritages, their provisions, and all their possessions. O Nut, cause the king to be restored that he may live" (Allen and Manuelian 2005, Faulkner 2007).

A male god associated with bees was Min (Figure 11.3), who was the god of male sexual potency and fertility. Like Neith and Nut, Min also goes back to the earliest period in Egypt's recorded history. The image of Min is rather startling to Western sensibilities. At Karnak Temple (and in general),

Figure 11.3 The god Min at Luxor Temple. Photograph by Gene Kritsky.

he is represented as a standing mummiform male wearing a double-feathered crown, with a prominent erection and an upraised right hand holding a flail (Romanosky 2001).

The apicultural relationship to Min is revealed in carved reliefs that specifically describe offerings of honey to Min. There were priests of Min who were temple beekeepers, and other priests may have collected wild honey. Temple ceremonies involved pouring thick honey over a statue of Min. The honey-offering reliefs found at Edfu and Dendera temples, described in Chapter 6, are accompanied by inscriptions that are rather sexual in nature—not unexpectedly, given their association with Min. The relief on the third column of the second row on the east side of the outer hypostyle hall at Edfu shows that the Egyptians considered honey an aphrodisiac and believed that it aided in fertilization (Zecchi 1997).

Honey bees also play a role in ancient Egypt's funerary literature. There were several such books in Egyptian theological history, but the Book of Going Forth by Day, or the Book of the Dead, is the best known. It contains some 200 spells that provided the deceased with the knowledge and magical power to survive the journey through the netherworld to the final afterlife (Lesko 2001). Other funerary books, such as the Book of That Which Is in the Underworld (Amduat) and the Book of Gates, describe the passage of Re's solar barque through the hours of the night. These "Underworld Books" and three sarcophagi that provide additional text divide the night into twelve hours, of which the Eighth Hour is of apian interest (Manassa 2008).

The Eighth Hour is divided into three registers. Re's barque proceeds through the middle register with human assistance. However, the upper and lower registers are divided into five caverns separated by doors. Each cavern is accompanied by a text that names the cavern and provides information about what sound is heard when Re enters the cavern. When Re enters the First Cavern of the Eighth Hour, he hears the sound of a swarm of bees. Manassa (2008) suggests that the sound of the swarm is to recall the bees' function as a creative force.

The events and sounds of the Second and Fourth Caverns of the Eighth Hour tie honey and bees to the cow goddess Hathor. She is a predynastic god, often represented as a woman with cow's ears (Figure 11.4), who was associated with music, love, beauty, and the "pleasurable" aspects of sex (Shaw and Nicholson 1995). Both the Second and Fourth Caverns include references to honey, along with the sounds of cymbals in the Second Cavern and the sound of a "bull being pleased" in the Fourth Cavern. This is reminiscent of a passage in the Mut Ritual, translated as "We should harmonize with the bees after the bull has cried out in the night" (Manassa 2008).

Hathor had a complex relationship with Re. In one myth, she was the eye of Re and considered his daughter. She has also been thought of as the mother of Re, and as his mother, she carried him as the sun disk between her horns. She was also thought of as his wife, whom Re impregnates with himself, making her his mother again (Vischak 2001). This complex association with Re, the creator of bees, is likely why honey was forbidden to be eaten during one of Hathor's festivals at her temple at Dendera (Manassa 2008). Hathor and honey

Figure 11.4 The goddess Hathor at Deir el-Bahri. Photograph by Jessee Smith and used with permission.

are consubstantial, so eating honey during these festivals is the same as eating Hathor, which would be catastrophic.

There were other festivals in which honey played a more positive role, and one of these was the Opet Festival. This festival occurred during the second month of the Egyptian civil calendar at the time of the Nile inundation, when agricultural activities slowed (Spalinger 2001). At its origin, the Opet Festival was 11 days long, but it eventually lengthened to 27 days. The highlight of the festival was the pharaoh's appearance, leading a boat carrying statues of the gods Amun, Mut, and Khonsu down the Avenue of the Sphinxes from Karnak Temple to Luxor Temple and back again. As the festival evolved, the route changed from going over land to floating up and down the Nile. The purpose of the festival was to celebrate the marriage of Amun to Mut, and to celebrate the Egyptian state, reflecting Amun's role as the protector of the nation and the pharaoh. Coinciding with the annual flooding of the Nile made the Opet Festival a celebration of renewal. This was reflected in the "renewal of Amun in the person of his ever-renewing human vessel," the pharaoh. During the reign of the pharaoh Horemheb, the Opet Festival reconfirmed this association with the god Amun by the annual coronation of the king, who was the representative of Amun on Earth (Darnell 2010).

Feasting was part of this celebration, and at both Luxor and Karnak temples, there are registers (offering lists) that describe the menu of the festival. These are easily recognized by their characteristic grid of up to 40 sections describing the quantities of foods that were included in the festival. At Karnak Temple, the menu can be found on the upper part of the east half of the north wall facing the hypostyle columns. The relief shows the pharaoh Seti I on the left, facing a register of foods. The fourth section of the upper register (the one section with the honey bee hieroglyph) reads, "O Amun, take to thyself the Eye of Horus which is sweet to thy heart, white (?) honey, one vessel" (Nelson 1949a). A similar register at Luxor Temple in the sanctuary, one of the small rooms past the Court of Amenhotep and the hypostyle hall, features Amenhotep III presenting the menu for the festival (Figure 11.5). On the top row, the

Figure 11.5 Amenhotep III presenting the menu for the Opet Festival as seen in Luxor Temple. The offering of honey is shown in the highlighted box in the center of the top register. Photograph by Gene Kritsky (see color plate 12).

Figure 11.6 A register listing the food to be provided to Menna in the afterlife. Honey is in the lower far left. Photograph by Gene Kritsky (see color plate 13).

eighth section from the left (with the honey bee hieroglyph) lists two vessels of honey (Nelson 1949).

These registers listing foods for the deceased are also found in many tombs, and several also include honey (Cauville 2012). An exceptional grid list is easily found in the tomb of Menna (Figure 11.6), where we have already documented vessels of honey being carried into the tomb as part of a procession or included in stacks of food offerings. The register in Menna's tomb is painted rather than carved and is smaller than the lists in Karnak and Luxor temples, containing only 19 items rather than the 39 and 40 found in the temples. Menna's listing for one vessel of honey is found on the far left of the lower register in the grid. The top row of the grid, from left to right, includes three kinds of bread, meat, roast meat, wine, beer in a special jug, and another offering of beer, and the last square of the grid represents the transfiguration ritual. The bottom row, from left to right, includes honey, red water, natron, milk, roast meat, two other kinds of meat, two kinds of biscuits, and water in a particular jug. The strokes at the bottom of each section are the number of portions of each offering. (Hartwig 2013). These lists are essentially spells to ensure that the deceased is provided for in the afterlife. Similar grids are present in the tombs of Rekhmire and Nakht. Like Menna's listing, they include one vessel of honey in the same section of the grid, the far left of the lower register (Manniche 1986, Newberry 1900).

The ancient Egyptians provided physical offerings to their dead throughout their history. Early predynastic graves often contained pottery vessels alongside the deceased, who had been carefully laid out in a fetal position. The nearly intact tomb of Tutankhamun was the other extreme in this practice, with the tomb containing lavish furniture, jewelry, foodstuffs, chariots, beds, and other items felt necessary for the afterlife—including, of course, an amphora of honey (Carter number 614 j, pottery vessel, Griffith Institute 2004).

CHAPTER 12

The Magic of Beeswax

To the ancient Egyptians, beeswax seemed to have magical properties. Beeswax, along with honey, is mentioned in the Salt 825 papyrus as resulting from the work of the bee, a product from the tears of Re. This gift from the god was used in religion, medicine, and material arts as an adhesive, an embalming agent, a light source, and a medium for creating some of the most beautiful art of that time (Lucas 1948, Raven 1983).

Beeswax is produced by bees from glands found on the underside of the bee's abdomen. Wax-producing bees consume honey and bee bread (pollen that has been collected by bees and fermented in the hive) to obtain the raw materials that are metabolized to make the wax. Wax, when first secreted by the bee, is a clear liquid that quickly cools into a white flake or scale. The plasticity of wax allows the bees to use their mandibles to manipulate it to make comb: hexagonal cells in which the queen lays eggs, worker bees place nectar to be processed into honey, and pollen is stored for later consumption (Shimanuki et al. 2007).

Beeswax is a lipid, an organic substance that is insoluble in water. At low temperatures it is brittle, but it can be molded and shaped into various forms at higher temperatures and becomes fully liquid between 143° and 151° F/62° and 66° C (Shimanuki et al. 2007). The range in melting temperatures reflects the inclusion of different amounts of propolis (resin that is collected by bees and used to waterproof the hive).

The magical properties of beeswax relate to its physical properties. It is not affected by water, it does not discolor, and it will not lose its shape. It can be formed into figurines that will last for centuries. However, when it is placed in hot sunlight, its surface can change, which must have been an important quality to the Egyptians, considering their solar theology. Finally, beeswax burns. As it burns, it gives off a bright light, and it does not leave any ash (Raven 1983).

None of the archaeological beekeeping reliefs show how wax was harvested. In the tomb of Rekhmire, we do see white round honeycombs being collected from the hives, and we see a mound of a crushed white substance in a container, which is probably crushed honeycomb. However, we do not see any other step in the collection of beeswax.

White honeycomb is the purest comb in the hive, and even today it is prized as a sign of purity when sold as comb honey. The darker comb found in hives is brood comb, reused by the bees to produce more bees. Thus, the white round honeycombs depicted in Rekhmire's relief represent pure, newly harvested honeycombs. The next step of honey processing would be to separate the honey from the comb, and the Rekhmire relief suggests that the honey was allowed to simply separate from the crushed wax. That would still leave some honey adhering to the wax fragments, requiring more processing.

Columella, a Roman writer who lived between 4 and 70 CE, described the rendering process that was being followed at the beginning of the Roman rule of Egypt in 30 BCE. The process involved hot water and was useful when there were large amounts of wax that needed to be separated from honey and other contaminants. It was also much safer, as wax combined with water greatly reduces its fire hazard. According to Columella's description, the crushed beeswax/honey material was placed into a large container that was partially filled with water. The water was brought to a boil and then strained through straw or rushes to remove any particulate matter, such as parts of bees, before being boiled again. As the water cooled, the wax would float to the surface to be skimmed off (Serpico and White 2000). Generally, the more slowly the water cooled, the higher was the quality of the collected wax (Cowan 1908). Traditional Egyptian beekeepers in the twentieth century poured the wax/water liquid into bags for straining. In East Africa, traditional beekeepers using horizontal log hives also use bags to strain out impurities as they pour the hot water/wax mixture into a cold-water bath. As the water cools, the wax simply floats to the surface to be collected. I tested this by diluting crushed comb honey with water, bringing it to a boil, and then letting it slowly cool. As it cooled, clean white beeswax formed a skin on the surface, which was easily skimmed off.

How the ancient Egyptians strained the boiled water, if they used Columella's process, is unknown, as there are no reliefs that show this step. There are carved reliefs that show workers using cloth to squeeze out the last bits of grape juice to make wine, but my experiments using a low-thread-count cloth (a much lower thread count than typical linen from ancient Egypt) to squeeze out the honey from the wax/honey mixture proved unsuccessful.

Once processed, beeswax could be used for exchange. An ostracon in the Cairo Museum mentions 2 *deben* of wax, indicating that was how it was measured. In one instance, 60 *deben* of wax was valued at 30 *deben* of copper, so 1 *deben* of wax would be valued at half a *deben* of copper (Janssen 1975).

Pure, clean beeswax provided the Egyptians with a processed material that they used in a number of ways. The most obvious was as a source of light. The ubiquitous light source in ancient Egypt was the oil lamp, a small dish of oil into which a wick would be placed and lit. Such dishes have been found in Egypt, and the wick is even a hieroglyphic symbol. However, there are two reliefs that show "candles" being used (Murray 2004), including one candle that appears to be a wickless coil of beeswax (Crane 1999).

A second use of wax was as an adhesive. The worked pieces of flint that made up a sickle were "glued" into the handle with beeswax. In the Middle Kingdom, beeswax was combined with pulverized limestone to make a cement that was used to affix a razor to a handle (Lucas 1948).

The art of lost-wax casting, discussed in Chapter 1, was used in Egypt to cast solid copper objects during the Old Kingdom. Hollow objects were cast in copper during the Middle Kingdom, and by the New Kingdom, the process was used to create spectacular objects (Brier and Hobbs 2009, Ogden 2000). The cobra and the vulture on the funerary mask of Pharaoh Tutankhamun (Figure 12.1) were first carved in wax and used to create a mold that was used in casting these solid gold figures (Venable 2011). Other items cast in gold included massive signet rings, sections of gold pectorals, and statues, including the small gold statue of Amenhotep III (Hunt 1980). Leek (1975) published a photograph of a small, intricately carved beeswax model of Horus, which he believed was to be used in casting a golden statue.

The water-repelling properties of beeswax made it ideal for surface finishing, and it was logically used to waterproof Egyptian boats. By 300 BCE, it was mixed with paint and used to decorate the pharaoh's ships. Callisthenes reported that entire boats were covered with wax paintings (Crane 1999). Beeswax was also used on paintings in several tombs. In some instances, it was mixed with a pigment and used as a binder, or applied to the surface as a protective finish. Petrie reported that beeswax filled in some of the incised hieroglyphs on the granite sarcophagus of Ramesses III. It was also used as a coating to protect yellow pigment that filled in an incised inscription on a wooden box found in Tutankhamun's tomb (Lucas 1948).

Visitors to the Metropolitan Museum of Art, the Louvre, the Brooklyn Museum, the British Museum, and the Petrie Museum at the University College of London marvel at the vivid colors and fine detail of the Faiyum portraits. These strikingly lifelike images were found in a cemetery in the Faiyum that dates back to the first centuries of the Roman period. They were produced using the process of encaustic painting, which involved mixing pigments with hot or cold beeswax, egg, and linseed oil. This pigment mixture could then be applied with a brush in much the same way that oil paintings are created today. This permitted the artists to create painterly, expressive portraits with rich, lasting colors. One of the best examples is the portrait of the boy Eutyches, dating back to the second century, on display at the Metropolitan Museum of Art in New York (Metropolitan Museum of Art 2015a). A close examination

Figure 12.1 The funerary mask of Tutankhamun. The cobra and the vulture were made using the lost-wax process.

of the portrait of "Petrie's Red Youth" (Figure 12.2) (UC19610) at the Petrie Museum in London shows how the beeswax–pigment combination provided the artist with a medium that could capture the texture of individual brushstrokes to give added detail to the facial features (Figure 12.3).

Encaustic painting was brought into Egypt during the Ptolemaic Era as part of the Greek influence. It is thought that these portraits were initially created to hang in the home of the sitter. After the individual died, the portrait, which was painted on a wood panel, would be incorporated into the wrapped mummy and buried.

The care that was taken in producing the Faiyum portraits suggests that the sitters had some concern for their appearance, and beeswax played a role in that as well. A recipe from the Ebers papyrus describes a topical cream used to treat wrinkles, made from gum of frankincense, moringa oil, cyperus grass, and beeswax. These ingredients were ground, and then mixed with fermented plant juices. The resulting cream was applied daily (Manniche 1989).

Figure 12.2 "Petrie's Red Youth." From Petrie (1913).

Beeswax was also used to control unruly wigs. The ancient Egyptians made their wigs from human hair and, in some cases, incorporated plant fibers. The wigs were styled with small corkscrew curls and arranged with long plaits at the back. Lucas (1948) examined several wigs of human hair and plant fiber and found that beeswax was used to maintain the shape of the curls and to control the plaits. The wax would have been heated to the melting point and used to style the wig, and the curls would have been set once the wax cooled.

Beeswax also had a role in the afterlife. Some female mummies from the Eleventh and Eighteenth Dynasties were discovered with a thin 1- to 2-millimeter layer of beeswax on the face, along the back, and on the thighs. Toward the end of the New Kingdom, it was common use wax to plug or cover openings on the face.

Of all the ancient Egyptian uses for beeswax, the ones held in the highest regard involved its magical powers. The Salt 825 papyrus, the same document that told how bees were created, also included details of using wax for magic.

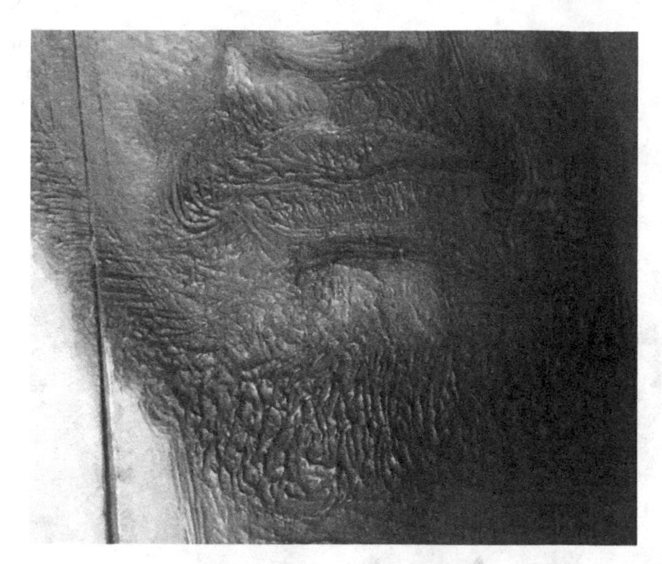

Figure 12.3 A close-up of "Petrie's Red Youth" showing the textured painting made possible by the encaustic technique, in which pigment was mixed with beeswax. Photograph by Gene Kritsky, and published courtesy of the Petrie Museum of Egyptian Archaeology, UCL (see color plate 14).

It describes how wax could be used to ensure the destruction of Seth, the god of confusion, disorder, and violence, and the murderer of Osiris (Velde 2001). To defeat Seth, the papyrus directed the making of beeswax figures of enemies to "kill the name of Seth" (Raven 1983). The destruction of the enemies was accomplished by burning the wax figures. Wax's property of burning without leaving any trace of the figure had a special appeal to the Egyptians. This may also help us understand why these wax figures are so rare, because many of them were burned as part of a ritual.

The oldest known beeswax figures date to the Ninth and Eleventh Dynasties and are in the form of a human male and female; in this case, they were possibly effigies of the tomb's owner and thought to function as spare bodies for the deceased (Raven 2012). These figures were thought to have the same function as the shabtis, shawabtis, and ushabtis. (These terms are not interchangeable but rather relate to the time period and where they were found. The earliest forms are considered shabtis, models that would perform magical activities for the deceased [Spanel 2001].) The British Museum collection includes one of these wax figures that was found in the tomb of Kawit, a royal wife who lived during the Eleventh Dynasty. This wax figure was placed in a small wooden coffin that was lined with linen (Figure 12.4). Another Eleventh Dynasty wax shabti is on display at the Museum of Fine Arts in Boston.

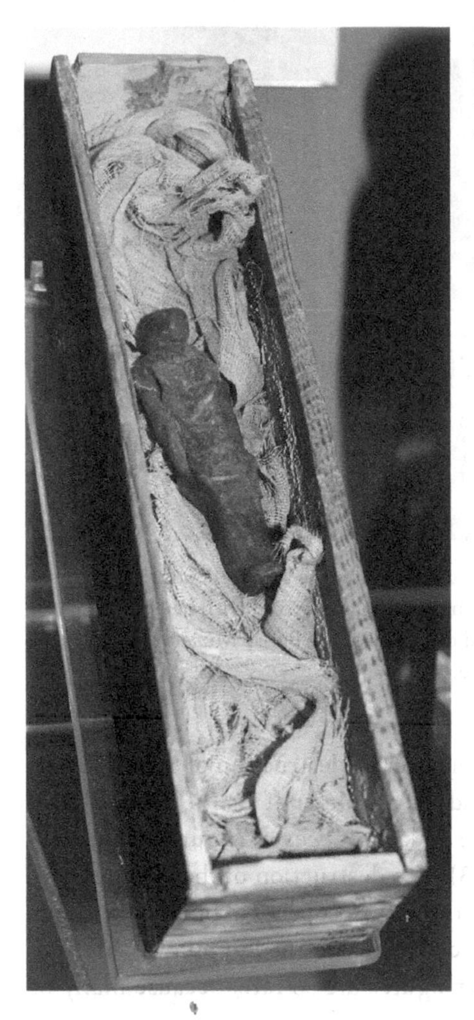

Figure 12.4 An Eleventh Dynasty beeswax shabti from the tomb of Kawit. Photograph by Gene Kritsky. © Trustees of the British Museum.

The Cleveland Museum of Art has two sets of wax figurines of the sons of Horus (Figures 12.5 and 12.6) that date to the Third Intermediate Period. The sons were Duamutef, Qebehsenuef, Imsety, and Hapy (although when wax figures were used it was common for an extra Imsety to be substituted for Qebehsenuef) and they protected the internal organs of the deceased. The gods' heads were often sculpted on the canopic jars that contained the stomach, intestines, liver, and lungs that had been removed during the process of mummification (Shaw and Nicholson 1995). The practice of using canopic jars stopped in the Twentieth Dynasty, when the internal organs were embalmed and returned to the abdominal cavity of the mummy. In the next dynasty, wax models of the four sons of Horus were included on the mummy. This defensive magic symbolically restored the protection of the gods (Raven 1983). Also at the Cleveland Museum of Art are two small beeswax models of facial masks

Figure 12.5 Four beeswax sons of Horus (from left to right): Hapy, Duamutef, Imsety, and Imsety. Hapy, 1000–900 BCE. Egypt, Third Intermediate Period, late Twenty-first Dynasty (1069–945 BCE) or early Twenty-second Dynasty (945–715 BCE). Wax with dark amber varnish; 8.2 × 2.2 cm. The Cleveland Museum of Art, Gift of the John Huntington Art and Polytechnic Trust 1914.690. Duamutef, 1000–900 BCE. Egypt, Third Intermediate Period, late Twenty-first Dynasty (1069–945 BCE) or early Twenty-second Dynasty (945–715 BCE). Wax with dark amber varnish; 9.3 × 2.0 cm. The Cleveland Museum of Art, Gift of the John Huntington Art and Polytechnic Trust 1914.691. Imsety, 1000–900 BCE. Egypt, Third Intermediate Period, late Twenty-first Dynasty (1069–945 BCE) or early Twenty-second Dynasty (945–715 BCE). Wax with dark amber varnish; 8.9 × 2.0 cm. The Cleveland Museum of Art, Gift of the John Huntington Art and Polytechnic Trust 1914.692. Imsety, 1000–900 BCE. Egypt, Third Intermediate Period, late Twenty-first Dynasty (1069–945 BCE) or early Twenty-second Dynasty (945–715 BCE). Wax with dark amber varnish; 8.6 × 2.0 cm. The Cleveland Museum of Art, Gift of the John Huntington Art and Polytechnic Trust 1914.693 (see color plate 15).

(Figure 12.7) that were attached to parts of mummies. Several beeswax amulets have also been found, including ankhs, offering tables, winged sun disks, tyets, and a collar (Petrie 1994).

Beeswax figurines of animals were also common. One example is the hippopotamus, which was the animal form of Seth. There are several texts, including the Salt 825 papyrus, that discuss the slaughtering of hippopotami by the destruction of the wax hippos. Birds modeled in wax include the ibis and

Figure 12.6 Four beeswax sons of Horus (from left to right): Hapy, Duamutef, Imsety, and Imsety. Hapy, 1000–900 BCE. Egypt, Third Intermediate Period, late Twenty-first Dynasty (1069–945 BCE) or early Twenty-second Dynasty (945–715 BCE). Honey-colored wax with dark amber varnish; 8.3 × 2.1 cm. The Cleveland Museum of Art, Gift of the John Huntington Art and Polytechnic Trust 1914.697. Duamutef, 1000–900 BCE. Egypt, Third Intermediate Period, late Twenty-first Dynasty (1069–945 BCE) or early Twenty-second Dynasty (945–715 BCE). Honey-colored wax with dark amber varnish; 8.8 × 2.1 cm. The Cleveland Museum of Art, Gift of the John Huntington Art and Polytechnic Trust 1914.696. Imsety, 1000–900 BCE. Egypt, Third Intermediate Period, late Twenty-first Dynasty (1069–945 BCE) or early Twenty-second Dynasty (945–715 BCE). Honey-colored wax with dark amber varnish; 8.2 × 2.0 cm. The Cleveland Museum of Art, Gift of the John Huntington Art and Polytechnic Trust 1914.694. Imsety, 1000–900 BCE. Egypt, Third Intermediate Period, late Twenty-first Dynasty (1069–945 BCE) or early Twenty-second Dynasty (945–715 BCE). Honey-colored wax with dark amber varnish; 8.1 × 2.0 cm. The Cleveland Museum of Art, Gift of the John Huntington Art and Polytechnic Trust 1914.695 (see color plate 16).

benu, or phoenix. There is a beautiful wax benu in the Cleveland Museum of Art collection (Figure 12.8). The benu models date from the Twenty-first and Twenty-second Dynasties and they were placed near the armpit of the mummy to ensure the resurrection of the deceased (Raven 2012). Wax scarabs may have also served as symbols of resurrection. Such a scarab was found between the linen wrappings on a mummy of a priest from Deir el-Bahri. Another scarab,

Figure 12.7 Mask, 332 BCE–395 CE. Egypt, probably Greco-Roman Period. Wax; 10.7 × 6.7 × 3.5 cm. The Cleveland Museum of Art, Gift of the John Huntington Art and Polytechnic Trust 1914.712.

this one in the collection at Leiden, was made of resin that was coated with a thin layer of wax inscribed with a spell from the Book of the Dead.

There are several spells or charms in the Greek magical papyri that rely on wax figures. Many of these are not tied to the afterlife but apparently were intended to help in the current life. One wax charm, modeled to look like a begging man with a bag and staff, helped attract business or served to "call in customers." Another charm, red wax modeled into a hippopotamus, was used to send dreams. There is also a spell to "bind a lover" that incorporated male and female wax figures (Betz 1986).

The magical properties of beeswax were part of Egyptian mythology dating back to the Twelfth Dynasty. One such myth, as recorded in the Westcar

Figure 12.8 Benu-bird, 1000–500 BCE. Egypt, Third Intermediate Period, late Twenty-first Dynasty (1069–945 BCE) or early Twenty-second Dynasty (945–715 BCE). Honey-colored wax with dark amber varnish; 3.2 cm. The Cleveland Museum of Art, Gift of the John Huntington Art and Polytechnic Trust 1914.689.

Papyrus, involves the priest Webaoner, who lived in the Third Dynasty. The priest fashioned a wax crocodile that was seven palms long, and he arranged to have it thrown into the pond where his wife's lover was bathing. Once in the water, the wax crocodile turned into a living crocodile that was 11½ feet in length. The crocodile seized the lover and disappeared. A week later, Webaoner, with the pharaoh as his witness, ordered the crocodile to show itself. Webaoner touched the crocodile, and it transformed back into the wax model, disgorging the lover. The pharaoh, upon seeing the crocodile with the lover, condemned the lover to his fate, and the crocodile transformed back into a real crocodile and was permitted to disappear with the lover forever (Raven 1983).

It is alleged that the Thirtieth Dynasty pharaoh Nectanebo was able to counter the Persian military by using spells combined with beeswax ships and men. Before a sea battle commenced, Nectanebo would go to a room, repeat spells, and act out the battle using wax models. After this ritual, his forces were victorious, but his magic could not always secure victory. During one of his battles, he saw the gods giving victory to the Persians, prompting him to secretly flee his palace (Raven 2012).

CHAPTER 13

The Afterlife of Ancient Egyptian Beekeeping

Egyptologists cite 394 CE as the last year of the ancient Egyptian culture, for that was the year the last hieroglyphic inscription was carved at Philae Temple. The crude hieroglyphs were inscribed just three years after Emperor Theodosius, a Christian, ordered the closing of all non-Christian temples. Few people at the time could still read the ancient symbols, and eventually the language ceased to be understood at all. Although the temples were closed, Rome still expected honey to be produced, and the ancient ways of Egyptian beekeeping survived not only the formal end of ancient Egypt but the end of the Roman Empire as well.

Bees and honey continued to have a special place in the hearts of early Egyptian Christians, and this encouraged beekeeping during these first centuries after the closing of the ancient temples. In 640 CE, an army led by Amr Ibn al-Aas conquered Egypt and it became part of the Islamic Empire. This religious and political change did not lessen the importance of bees and honey. Indeed, honey's importance is underscored in the Qu'ran's An-Nahl (The Bee); 68–69 reads:

> And the Lord taught the Bee to build its cells in hills, in trees, and in men's habitats; then to eat of all the produce and find with the skill the precious paths of its Lord: there issues from within their bodies a drink of varying colours, wherein is healing for men; verily this is a sign for those who give thought. (Sultanate of Oman 2002)

Beekeeping using traditional Egyptian techniques continued during the centuries that followed this conquest, and although we do not know all the details of these practices, it seems likely that some of the techniques used during pharaonic times survived this 1,700-year period.

Figure 13.1 A wall apiary south of Minya in central Egypt. Photograph by Gene Kritsky.

Traditional beekeepers in modern-day Egypt still rely on horizontal cylinder hives, reminiscent of those shown in the oldest relief of Newoserre Any's sun temple, though the hives are not tapered at the end (Figure 13.1). These mud hives are made using a mat constructed from reeds or other straight stalks that are stitched together. The mat is laid flat on the ground and covered with a thick layer of mud, clay, and small pieces of plant material. Once the mat is completely covered, it is rolled up around a core of bound sticks and tied into a tube, with the mud layer between the core and the outer mat. This is allowed to dry in the hot Egyptian sun for one to two weeks. Once dry, the core is pushed out of the center, the mat is removed, and the tubular hive is ready for use (Crane 1999). When completed, hives are between 120 and 140 centimeters long and approximately 25 centimeters in outside diameter, with an internal diameter between 17 and 22 centimeters. Disks of mud molded by hand and dried in the sun are used as end caps for the hives (Figure 13.2).

A wall of hives is constructed by lining up the hive tubes parallel to one another on the ground, with the sidewalls touching. A second layer of hives is placed on the first layer, with each hive fitting into the space between two lower hives (Figure 13.3). Any gaps between the hives are filled in with straw and mud (Figure 13.4). This process continues until the wall apiary reaches the desired size. Many of these walls consist of approximately 400 hives, but some could be even larger. I visited a wall apiary south of Minya that was composed

Figure 13.2 Mud end caps for traditional Egyptian beehives drying in the sun. Photograph by Gene Kritsky.

of 106 hives along the ground, with rows stacked 10 hives high. Many of the hives on the bottom and top were empty, but there were well over 800 working hives in this wall.

In traditional beekeeping, bees are obtained from established hives. Working from the rear of the hive, the beekeeper smokes the colony to crowd the bees toward the front of the hive. He then calls the bees by making a *kak,*

Figure 13.3 Stacked traditional horizontal beehives, before spaces were filled in with mud. Photograph by Gene Kritsky.

Figure 13.4 Spaces between the stacked hives are filled in with mud to prevent bees from colonizing the spaces between the hives. Photograph by Gene Kritsky.

kak, kak sound near the hive's entrance. This call imitates the piping of virgin queens, to which a new queen in the hive or still in her queen cell will reply. If the beekeeper hears a queen return the call, he will remove three combs with brood and place them near the front of an empty hive. Each piece of comb is affixed to a stick that is slightly longer than the diameter of the tube, which props up and supports the comb within the hive. The beekeeper then crowds the bees toward the front using smoke and scoops many of them out of the hive with a ladle attached to a very long handle. If he does not catch the old queen in the scoop, he will make certain to include several virgin queens in the new hive. Many hives produce between 10 and 20 queen cells per hive. However, *Apis mellifera lamarckii*, the subspecies found in parts of Upper Egypt, may produce up to 250 queen cells, and many young queens might live together in a hive until the colony swarms (Crane 1999). After the bees have been transferred, the hive is marked with one to four white spots, which tell the beekeeper when the colony was started.

Archaeological reliefs show ancient Egyptian beekeepers using smoke from a censer to quiet the bees and a variety of pots to harvest and store honey. No other special beekeeping tools are known from the reliefs. Beekeepers today working horizontal mud hives use an arsenal of long-handled tools to gather the bees and manipulate the comb. The aforementioned ladle, called the *makah*, is used to collect bees and transfer them to another hive. Another common tool is the *sadif*, a flat-bladed, spatulate tool at one end of a 120-centimeter handle (Figure 13.5). The *sadif* blade is about 10 centimeters long and 3 centimeters wide at the base; it tapers to 2 centimeters wide at the tip. It

Figure 13.5 The end of a *sadif*, a flat-bladed tool used to help remove comb from traditional Egyptian hives. Photograph by Gene Kritsky.

is used to cut along the edge of the circular comb to free it from the walls of the hive. Another long-handled tool, the *showka* (Figure 13.6), is furnished with a short, flat sickle at one end, which is used to remove queen cells and to help manipulate comb after cutting it from the hive walls. At the other end of one tool that I examined was a cleaning tool called the *attala*, which resembled the curved end of the Western hive tool or a paint scraper, and is used to scrape comb debris from inside the hive (Crane 1999).

Traditional beekeepers are quite adept in using these tools, sometimes wielding one in each hand. I have observed a beekeeper removing the end cap of a hive, carefully using the *sadif* to cut along the sides of the comb where the bees had attached it to the hive, and then holding the *sadif* in his left hand and the *showka* in his right hand, with the shafts crossing in the middle like tongs (Figure 13.7). With the crossed tools, he carefully moved the comb to the rear

Figure 13.6 The *showka* has a flat sickle-shaped blade at one and a cleaning tool, called the *attala*, at the other. Photograph by Gene Kritsky.

Figure 13.7 A beekeeper using two long tools to remove a comb from a traditional hive. Photograph by Gene Kritsky.

of the hive, where he could remove it with his hands to harvest it (Figure 13.8), if it was sealed honey, or to place it into another hive if it was mostly sealed brood. Generally, honey was found toward the rear of the hive and brood comb was in the front.

Honey was harvested by removing the sealed comb from the hives and then extracting by hand by pushing the comb through *vulu*, the husk-like bases of date palm leaves. There are three honey seasons in Egypt: citrus honey is harvested (mostly in Lower Egypt) in early April; clover honey is harvested in May and the first week of June; and cotton honey is collected from August through September (Hussein 2001).

The traditional horizontal tube hives were used not only in Egypt during the Common Era but also throughout the Middle East, including Israel, Palestine,

Figure 13.8 A beekeeper removing one of the round honeycombs from a traditional hive. Photograph by Gene Kritsky.

Jordan, Syria, Turkey, Iraq, Iran, and the Arabian Peninsula (Crane 1999). Hasselquist (1766: 153–154) visited Palestine in 1751 and reported, "The inhabitants breed a great number of bees, to their considerable advantage, and with little trouble. They make their bee-hives out of clay, four feet long and half a foot in diameter as in Egypt: they lay ten or twelve of them on one another on the bare ground, without anything under them."

Traditional Egyptian beekeepers also practiced migratory beekeeping in the 17th and 18th centuries (Newberry 1938). As described by C. Pococke (1743: 210):

> If I was rightly inform'd, they have an extraordinary custom in relation to their bees in Upper Egypt. They load the boats with the hives, at a time when their honey is spent; they sail down the stream all night, and take care to stop in a place by day, where the diligent animal may have the opportunity of collecting its honey and wax; and so, making a voyage of six weeks or two months, they arrive at Cairo, with plenty of honey and wax, and find a good market for both.

The number of hives moved was staggering. In 1760, some 20 boats reportedly moved 4,000 hives (Crane 1999). Twenty-seven years later, C. E. Savery (1787: 207–208) recorded the following on migratory beekeeping:

> Their manner of raising bees is no less extraordinary, and bespeaks great ingenuity. Upper Egypt, preserving its verdure only four or five months, the flowers and harvests being seen no longer, the people of the lower Egypt profit by this circumstance, assembling on board large boats the bees of different villages. Each proprietor confides his hives, with his own mark, to the boatman; who, when loaded, gently proceeds up the river, and stops at every place where he finds verdure and flowers. The bees swarm from their cells, at break of day, and collect their nectar, returning, several times, loaded with booty, and, in the evening, re-enter their hives, without ever mistaking their abode. Thus sojourning three months on the Nile, the bees, having extracted the perfumes of the orange flowers of the Said, the essence of the roses of Fayoum [*sic*], the sweets of the Arabian jasmine, and of every flower, are brought back to their homes, where they find new riches. Thus do the Egyptians procure delicious honey, and plenty of wax. The proprietors pay the boatmen, on their return, according to the number of the hives which they have taken from one end of Egypt to the other.

Newberry (1938: 32) reported that he observed a similar scene a century later. He wrote:

> In the winter of 1890 I remember seeing on the Nile, near Abu Korkas in Middle Egypt, a boat laden with well-stocked cylindrical beehives, which were being taken out and carried to the bean-fields in the neighbourhood, where they were left for the flowering season; after a few weeks they were brought back to the river, packed again on the boat, and transported northwards to other pasture grounds. By the early summer the boatmen told me, they were returned to their owners, who lived in the Delta.

It is possible that this form of migratory beekeeping was practiced during pharaonic times. As discussed earlier, beehives were leased out during the Ptolemaic Dynasty, and hives were moved by donkeys. Also, Pliny (23 CE–79 CE), who was a contemporary of the Roman rule of Egypt, reported that bees were moved by boat in Italy. However, there is no definitive evidence that the ancient Egyptians used boats to move their bees (Crane 1999).

In addition to migratory beekeeping, traditional mud hives also were used in urban beekeeping in Cairo during the 19th century. Keimer (1957: 24) described a European visitor who wrote that in Cairo, "there are those who follow no other trade except that of raising bees. They have up to a hundred hives on their roof-tops. The bees go out into the gardens and fields and come back without fail." He quoted another visitor who wrote, "On the roofs of some houses one also notices earthen cylinders that serve as beehives, not only around Cairo, but within the town itself."

Traditional Egyptian beekeeping, after being practiced for nearly five millennia, is on the wane. The first movable-frame hives were introduced into Egypt in 1880. These hives were populated with *Apis mellifera cypria*, as opposed to the local Egyptian bee, *Apis mellifera lamarckii*. By 1900, the modern box hive was common. In 1912, the Egyptian Ministry of Agriculture started to promote the new methods of beekeeping and using different subspecies of honey bees. The topography of Egypt made it ideal for raising pure strains of bees, as apiaries could be geographically isolated from each other. At first, the Cyprian bee was favored, but by the 1930s, Carnolian and Italian honey bees were being bred as efforts to improve the Egyptian bee were abandoned. Today, some beekeepers are trying to revive the Egyptian bee, which is thought to have survived in some of the apiaries in Upper Egypt, but the Carnolian bee remains popular. A small Kenyan top-bar hive has recently been introduced to better manage the Egyptian bee for this revival, rather than the cylindrical mud hive (Page et al. 1981).

Figure 13.9 Modern box hives with movable frames are quickly becoming more common than the traditional hives. Photograph by Gene Kritsky.

Many Egyptian beekeepers are using boxes in much the same way they kept bees in the traditional hives. Most of the box hives are maintained as single boxes without a full set of frames. If a beekeeper wishes to increase his production, he does so by increasing the number of single boxes (Figure 13.9) rather than increasing the population of a hive by supering (stacking new hive boxes on top of existing ones) (Abou-Shaara 2009).

Beekeeping's popularity waned during and immediately following the Second World War but was revived with the importation of 7,000 queen bees in 1953. These bees were to be the foundation of an outcrossing campaign with local bees. This influx of new bees coincided with the decline in traditional beekeeping. In 1926, there were fewer than 300 movable-frame hives in Egypt. By 1952, there were 16,000 modern hives, and by 1979, the number had exploded to 521,000 (Page et al. 1981).

This transition to new beekeeping methods was not without problems. In 1977, mites became a serious problem in Egypt and honey yields started to decline. Every box hive that I examined in 1981 was heavily infested with *Varroa* mites (Figure 13.10). Beekeepers were and still are concerned about the impact the heavy use of pesticides is having on bee populations.

The push to modernize beekeeping has continued over the past 20 years. In 1994, there were 1,119,000 modern box hives in Egypt, compared to 124,000 traditional mud hives (Hussein 2000). By 2009, the number of traditional mud

Figure 13.10 Modern beekeeping has brought modern problems. This hive was heavily infested with *Varroa* mites. Photograph by Gene Kritsky.

hives dropped to 7,700, while box hives increased to 1,344,000 (Abou-Shaara 2009). Once the predominant hive in Egypt, cylindrical mud hives account for less than 1% of the hives in use today. An ancient form of beekeeping that was practiced when the pyramids were built, outlasted the internal upheaval of the First Intermediate Period, survived waves of external invaders, expanded during the splendor of the New Kingdom, and endured beyond the Roman Empire has nearly disappeared in just half a century. When these traditional hives are no longer being used, this apicultural connection to ancient Egypt, and to the original keepers of the tears of Re, will be broken.

LITERATURE CITED

Abou-Shaara, H. F. 2009. Egyptian beekeeping. Serbian Beekeeping Journal. http://pcela.rs/Egyptian_Beekeeping_1.htm.

Allen, J. P., and P. D. Manuelian. 2005. The ancient Egyptian pyramid texts. Society of Biblical Literature, Atlanta.

Altenmüller, H. 2001. Old Kingdom. In Redford, D. B. The Oxford encyclopedia of ancient Egypt. Oxford University Press, New York. 2: 585–605.

Andrews, C. 1981. The British Museum book of the Rosetta stone. Peter Bedrick Books, New York.

Baedeker, K. 1902. Egypt; handbook for travellers. Karl Baedeker, Publisher; Leipsic.

Benderitter, T. 2007. Benia TT343. http://www.osirisnet.net/tombes/nobles/benia/e_benia_01.htm.

Benderitter, T. 2008. Userhat TT56. http://www.osirisnet.net/tombes/nobles/ous56/e_ouserhat56_01.htm.

Berenbaum,. M. 2010. Honey, I'm homemade: sweet treats from the beehive across the centuries and around the world. University of Illinois Press, Urbana.

Berman, L. M. 1999. The catalog of Egyptian art: the Cleveland Museum of Art. Cleveland Museum of Art, Cleveland, OH.

Betz, H. D. 1986. The Greek magical papyri in translation. University of Chicago Press, Chicago.

Bingen, J. 2007. Hellenistic Egypt; monarchy, society, economy, culture. University of California Press, Berkeley.

Bleiberg, E. 2001. Prices and payment. In Redford, D. B. The Oxford encyclopedia of ancient Egypt. Oxford University Press, New York. 3:65–68.

Borchardt, L. 1905. Das Re-Heiligtum des Königs Ne-woser-re (Rathures). Bd.1 Der Bau. A. Duncker, Berlin.

Breasted, J. H. 1906. Historical documents. Volume II. University of Chicago Press, Chicago.

Breasted, J. H. 1906a. Historical documents. Volume III. University of Chicago Press, Chicago.

Breasted, J. H. 1906b. Historical documents. Volume IV. University of Chicago Press, Chicago.

Breasted, J. H. 1916. Ancient times: a history of the early world. Ginn, Boston.

Brewer, D. J., D. B. Redford, and S. Redford. 1994. Domestic plants and animals: the Egyptian origins. Aris and Phillips, Warminster, England.

Brewer, D. J., and E. Teeter. 2007. Egypt and the Egyptians. Cambridge University Press, Cambridge, United Kingdom.

Brier, B., and H. Hobbs. 2009. Ancient Egypt: everyday life in the land of the Nile. Sterling, New York.

Brinton, J. 1979. Henry Salt, Esq. History Today 29(6): 358–365.

British Museum. 2013. Collection online, EA30550. http://www.britishmuseum.org/research/collection_online/collection_object_details.aspx?objectId=170297&partId=1&searchText=30550&page=1.

Budge, E. W. B. 1988. From fetish to god in ancient Egypt. Courier Dover, New York.

Cauville, S. 2012. Offerings to the gods in Egyptian temples. Peeters, Leuven, Walpole, MA.

"City of 20,000 Dead Found at Sakkara." 1938. New York Times, July 10, 1938, p.16.

Clayton, P. A. 1994. Chronicle of the pharaohs. Thames and Hudson, London.

Cowan, T. W. 1908. Wax craft. Sampson Low, Marsten, London.

Crane, E. 1983. The archaeology of beekeeping. Cornell University Press, Ithaca, NY.

Crane, E. 1999. The world history of beekeeping and honey hunting. Routledge, New York.

Darby, W. J., P. Ghalioungui, and L. Grivetti. 1977. Food: the gift of Osiris. Academic Press, London. Vol. 1.

Darnell, J. 2010. Opet festival. UCLA Encyclopedia of Egyptology. http://escholarship.org/uc/item/4739r3fr.

David, R. 1986. The pyramid builders of ancient Egypt. Routledge, London.

Davies, N. 1917. The tomb of Nakht at Thebes. Metropolitan Museum of Art, New York.

Davies, N. 1943. The tomb of Rekh-mi-re at Thebes. Metropolitan Museum of Art, New York.

Davies, N. M. 1936. Ancient Egyptian paintings. Vol. I, Plates I–LII. University of Chicago Press, Chicago.

De Jong, A. 2001. Feline deities. In Redford, D. B. The Oxford encyclopedia of ancient Egypt. Oxford University Press, New York. 1: 512–513.

Department of Ancient Near Eastern Art. 2004. "The Nahal Mishmar Treasure." In Heilbrunn Timeline of Art History. New York: Metropolitan Museum of Art, 2000–. http://www.metmuseum.org/toah/hd/nahl/hd_nahl.htm.

Derchain, P. 1965. Le Papyrus Salt (BM 10051), ritual pour la conservation de la vie en Egypte. Académie de Belgique Mémoires 58 fascicule 1a, Bruxelles.

Dodson, A., and S. Ikram. 2008. The Tomb in ancient Egypt. Thames and Hudson, London.

Edel, E. 1974. Die Jahreszeitenrelifs aus dem Sonnenheiligtum des Königs Ne-user-Re. Academie-Verlag, Berlin.

Eisenberg, J. M. 2002. Two new tombs found at Saqqara. Minerva 13(6): 7.

Emanuel, J. P. 2013. "ŠRDN from the sea": the arrival, integration, and acculturation of a "sea-people." Journal of Ancient Egyptian Interconnections 5(1): 14–27.

Estes, J. W. 1993. The medical skills of ancient Egypt. Science History Publications, Canton, MA.

Eyre, C. J. 1998. The market women of pharaonic Egypt. In Grimal, N. and B. Menu, Le commerce en Égypt ancienne. Institut Français d'Archéologie Orientale. BdE 121, IFAO: 173–189, Cairo.

Faulkner, R. O. 2007. The ancient Egyptian pyramid texts. Digireads.com, Stilwell, KS.

Fischer, H. G. 1973. Offering stands from the pyramid of Amenemhet I. Metropolitan Museum Journal 7:123–126.

Franke, D. 2001. First intermediate period. In Redford, D. B. The Oxford encyclopedia of ancient Egypt. Oxford University Press, New York. 1: 526–532.

Frankfurter, D. 1998. Religion in Roman Egypt: assimilation and resistance. Princeton University Press, Princeton, NJ.

García, J. C. M. 2010. La gestion des aires maginales: *phw, gs, tnw, sht*, au IIIe millenaire. In Woods, A., A. McFarlane, and S. Binder. Egyptian culture and society: studies in honour of Nabuib Kanawati. Publications du Conseil Suprême des Antiquités de l'egypte. II: 49–70.

Gardiner, A. H. 1916. The defeat of the Hyksos by Kamose: the Carnarvon tablet, No. 1. Journal of Egyptian Archaeology 3(2/3): 95–110.

Gardiner, A. H. 1988. Egyptian grammar. Griffith Institute Ashmolean Museum, Oxford, UK.

Grenfell, B. P., and H. S. Hunt. 1898. The Oxyrhynchus papyri, part I. Egypt Exploration Fund, London.

Griffith, F. L. 1927. The Abydos decree of Seti I at Nauri. Journal of Egyptian Archaeology 13: 193–208.

Griffith Institute. 2004. Tutankhamun: anatomy of an excavation. The Howard Carter archives. http://www.griffith.ox.ac.uk/gri/carter/614j-c614 j.html.

Haring, B. 2001. Temple administration. In Redford, D. B. The Oxford encyclopedia of ancient Egypt. Oxford University Press, New York. 1: 21–23.

Hartwig, M. 2013. The tomb chapel of Menna (TT69). American Research Center in Egypt Conservation Series 5, American University of Cairo. Cairo, Egypt.

Hassan, S. B. 1938. Excavations at Saqqara. Annales du Service des Antiquités de l'Egypte xxxviii, 503–521.

Hasselquist, F. 1766. Voyages and travels in the Levant in the years 1749, 50, 51, 52. L. Davis and C. Reymers, London.

Hein, I. 2003. A rediscovered ancient Egyptian offering stand at Ûzmir. Epigraphica Anatolica 35: 125–140.

Hirst, J. J. 2012. Menna—TT69. http://www.osirisnet.net/tombes/nobles/menna69/e_menna_02.htm.

Hodel-Hoenes, S. 2000. Life and death in ancient Egypt: scenes from private tombs in New Kingdom Thebes. Cornell University Press, Ithaca, NY.

Hornsey, I. S. 2012. Alcohol and its role in the evolution of human society. RSC Publishing, London.

Hunt, L. B. 1980. The long history of lost wax casting. Gold Bulletin 13: 63–81.

Hussein, M. H. 2000. A review of beekeeping in Arab countries. Bee World 81(2): 56–71.

Hussein, M. H. 2001. Beekeeping in Africa: I—North, East, North-east and West African countries. Proceedings of the 37th International Apicultural Congress, 28 October–1 November 2001, Durban, South Africa. http://www.apimondia.com/congresses/2001/Papers/001.pdf.

Ikram, S. 1994. Food for eternity: what the ancient Egyptians ate and drank. KMT 5(2): 53–60, 73.

Ikram, S. 2000. Meat processing. In Nicholson, P. T., and I. Shaw. Ancient Egyptian materials and technology. Cambridge University Press, Cambridge.

Janick, J. 2002. Ancient Egyptian agriculture and the origins of horticulture. 23–39. In Sansavini, S., and J. Janick, eds. Proceedings of the International Symposium on Mediterranean Horticulture Issues and Prospects. Acta Horticulturae 582.

Janssen, J. J. 1963. An unusual donation stela of the Twentieth Dynasty. Journal of Egyptian Archaeology 49: 64–70.

Janssen, J. J. 1975. Commodity prices from the Ramessid Peirod. E. J. Brill, Leiden.

Kadish, G. E. 1966. Old Kingdom Egyptian activity in Nubia: some reconsiderations. Journal of Egyptian Archaeology 52: 23–33.

Keimer, L. 1957. Bees and honey in ancient Egypt. Egypt Travel Magazine 30(February): 21–28.

Kritsky, G. 1993. Beetle gods, king bees, and other insects of ancient Egypt. KMT 4: 32–39.

Kritsky, G. 2010. The quest for the perfect hive. Oxford University Press, New York.

Kuény, G. 1950. Scenes apicoles dans l'ancienne Egypte. Journal of Near Eastern Studies 9(2): 84–93.

Labrousse, A., and A. Moussa. 2002. La chaussée du complexe funéraire du roi Ounas. Institut fançais d'archéologie orientale, Bibliothéque d'étude, Le Caire.

Lansing, A. 1920. The Egyptian expedition 1916–1919: II. Excavations in the Asasif at Thebes. Season of 1918–1919. Metropolitan Museum of Art Bulletin 15(7 pt. 2): 11–24.

Leek, F. F. 1967. The practice of dentistry in ancient Egypt. Journal of Egyptian Archaeology 53 (December): 51–58.

Leek, F. F. 1975. Some evidence of bees and honey in ancient Egypt. Bee World 56(4): 141–148, 163.

Lesko, L.H. 2001. Funerary literature. In Redford, D. B. The Oxford encyclopedia of ancient Egypt. Oxford University Press. 1: 570–575.

Lesko, L. H. 2001a. Nut. In Redford, D. B. The Oxford encyclopedia of ancient Egypt. Oxford University Press, New York. 2: 558–559.
Lucas, A. 1948. Ancient Egyptian materials and industries. Edward Arnold, London.
Manassa, C. 2008. Sounds of the netherworld. In Rothöhler, B. and A. Manisali, eds. Mythos and Ritual, Festschrift für Jan Assmann zum 70. Geburtstag. Lit Verlag, Münster, pp. 109–135.
Manley, D., and P. Rée. 2001. Henry Salt. Libri Publications, London.
Manniche, L. 1986. The tomb of Nakht, the gardener, at Thebes (No. 161) as copied by Robert Hay. Journal of Egyptian Archaeology 72: 55–78.
Manniche, L. 1989. An ancient Egyptian herbal. University of Texas Press, Austin.
Martin, G. T. 1971. Egyptian administrative and private-name seals principally of the Middle Kingdom and Second Intermediate Period. Ashmolean Museum, Oxford, UK.
Mazar, A., and N. Panitz-Cohen. 2007. It is the land of honey: beekeeping at Tel Rehov. Near Eastern Archaeology 70(4): 202–219.
McGovern, P. E. 2009. Uncorking the past. University of California Press, Berkeley.
Mehdawy, M., and A. Hussein. 2010. The Pharaoh's kitchen. American University in Cairo Press, Cairo, Egypt.
Metropolitan Museum of Art. 2015. Baskets of Honeycomb and Fruit, Tomb of Qenamun. The Collection Online, http://www.metmuseum.org/collection/the-collection-online/search/557703?rpp=30&pg=1&ft=honeycomb+Qenamun&pos=1.
Metropolitan Museum of Art. 2015a. Portrait of the boy Eutyches. The Collection Online, http://www.metmuseum.org/collection/the-collection-online/search/547951.
Müller, M. 2001. Re and Re-Horakhty. In Redford, D. B. Oxford encyclopedia of ancient Egypt. Oxford University Press, New York. 3: 123–126.
Müller, M. 2001a. Relief Sculpture. In Redford, D. B. Oxford encyclopedia of ancient Egypt. Oxford University Press, New York. 3: 132–139.
Murray, M. A. 2000. Fruits, vegetables, pulses, and condiments. In Nicholson, P. T., and I. Shaw. Ancient Egyptian materials and technology. Cambridge University Press, Cambridge, pp. 609–655.
Murray, M.A. 2004. The splendor that was Egypt. Dover Publications, Mineola, NY.
Museum of Reconstructions. 2003. Delight of Re: solar temple of Nuiserre. http://www.reconstructions.org/mor/pages/frames/mor_dor_splash/mor_dor_splash_frame.html.
Naunton, C. 2007. The tomb of Harwa (TT37)—a high official of the Twenty-fifth Dynasty. Ancient Egypt 7(3): 25–33.
Nelson, H. H. 1949. Certain reliefs at Karnak and Medinet Habu and the ritual of Amenophis I. Journal of Near Eastern Studies 8(3): 201–232.
Nelson, H. H. 1949a. Certain reliefs at Karnak and Medinet Habu and the ritual of Amenophis I—concluded. Journal of Near Eastern Studies 8(4): 301–345.
Newberry, P. E. 1900. The life of Rekhmara. Archibald Constable, London.
Newberry, P. E. 1905. Ancient Egyptian scarabs; an introduction to Egyptian seals and signet rings. Ares, Chicago.
Newberry, P. E. 1938. Bee-hives in Upper Egypt. Man 38: 32–32.
Nunn, J. F. 1996. Ancient Egyptian medicine. British Museum Press, London.
Ogden, J. 2000. Metals. In Nicholson, P. T., and I. Shaw, Ancient Egyptian materials and technology. Cambridge University Press, Cambridge, UK, pp. 148–176.
Oppenheim, A. 2011. The early life of pharaoh: divine birth and adolescence scenes in the causeway of Senwosret III at Dahshur. In Bárta, M., F. Coppens, and J. Krejãí. Abusir and Saqqara in the Year 2010/1. Czech Institute of Egyptology Charles University, Prague.
Oppenheim, A. 2013. Personal communication.
Page, R. E., M. M. Ibrahim, and H. H. Laidlaw. 1981. The history of modern beekeeping in Egypt. Gleanings in Bee Culture 109(1): 24–26.
Pardey, E. 2001. Provincial administration. In Redford, D. B. The Oxford encyclopedia of ancient Egypt. Oxford University Press, New York. 1: 16–20.

Parkinson, R. 2008. The painted tomb-chapel of Nebamun. British Museum, London.
Petrie, W. M. F. 1900. The royal tombs of the First Dynasty. Part I. The Egypt Exploration Fund, London.
Petrie, W. M. F. 1913. The Hawara Portfolio: Paintings of the Roman age. School of Archaeology in Egypt, University College and Bernard Quaritch, London.
Petrie, W. M. F. 1994. Amulets. Martin Press, London.
Petrie, W. M. F., and F. L. Griffith. 1898. Hieratic papyri from Kahun and Gurob: (principally of the Middle Kingdom). Bernard Quaritch, London.
Petrie, W. M. F., F. L. Griffith, and P. E. Newberry. 1890. Kahun, Gurob, and Hawara. Kegan Paul, Trench, Trübner, London.
Pococke, R. 1743. Description of the "east." W. Bowyer, London.
Poole, F. 2001. "Cumin, set milk, honey": an ancient Egyptian medicine container. Journal of Egyptian Archaeology 87: 175–180.
Posener-Kriéger, P. 1978. A letter to the governor of Elephantine. Journal of Egyptian Archaeology 64: 84–87.
Quibell, J. E. 1898. The Ramesseum. Egyptian Research Account 1896, Bernard Quaritch, London.
Quirke, S. 2001. State administration. In Redford, D. B. The Oxford encyclopedia of ancient Egypt. Oxford University Press, New York. I: 13–16.
Quirke, S. 2001a. Second Intermediate Period. In Redford, D. B. The Oxford encyclopedia of ancient Egypt. Oxford University Press, New York. 3: 260–265.
Ransome, H. 1986. The sacred bee in ancient times and folklore. Bee Books New and Old, Bridgwater, England.
Raven, M. J. 1983. Wax in Egyptian magic and symbolism. Oudheidkundige Mededelingen het Rijksmuseum van Oudheden te Leiden 64: 7–47.
Raven, M. J. 2012. Egyptian magic; the quest for Thoth's book of secrets. American University in Cairo Press, Cairo, Egypt.
Redford, D. B. 2001. The Oxford encyclopedia of ancient Egypt. Oxford University Press, New York.
Ritner, R. K. 2001. Magic in medicine. In Redford, D. B. The Oxford encyclopedia of ancient Egypt. Oxford University Press, New York. 2: 326–329.
Ritner, R. K. 2001a. Medicine. In Redford, D. B. The Oxford encyclopedia of ancient Egypt. Oxford University Press, New York. 2: 353–356.
Robins, G. 1986. Egyptian painting and relief. Shire Publications, Aylesbury, UK.
Robins, G. 1994. Proportion and style in ancient Egyptian art. University of Texas Press, Austin.
Robins, G. 2001. Grid systems. In Redford, D. B. The Oxford Encyclopedia of ancient Egypt. Oxford University Press, New York. 1: 68.
Romanosky, E. 2001. Min. In Redford, D. B. The Oxford encyclopedia of ancient Egypt. Oxford University Press, New York. 2: 413–415.
Root, A. I. 1888. The ABC of bee culture. A. I. Root, Medina, OH.
Root, A. I., and E. R. Root. 1923. The ABC and XYZ of bee culture. A. I. Root, Medina, OH.
Sagrillo, T. L. 2001. Bees and honey. In Redford, D. B. The Oxford encyclopedia of ancient Egypt. Oxford University Press, New York. 1: 172–174.
Säve-Söderbergh, T. 1957. Private tombs at Thebes, Volume 1, Four Eighteenth Dynasty tombs. Griffith Institute, Oxford.
Savery, C. E. 1787. Letters on Egypt. G. G. J. and J. Robinson, London. Vol. II.
Scott, N. E. 1952. Two statue groups of the V Dynasty. Metropolitan Museum of Art Bulletin, New Series, 11(4): 116–122.
Serpico, M., and R. White. 2000. Oil, fat, and wax. In Nicholson, P. T. and I. Shaw, Ancient Egyptian materials and technology. Cambridge University Press, Cambridge, pp. 390–429.
Shaw, I. 2002. The Oxford history of ancient Egypt. Oxford University Press, New York.

Shaw, I., and P. Nicholson. 1995. The dictionary of ancient Egypt. British Museum, Harry N. Abrams, New York.
Shimanuki, H., K. Flottum, and A. Harman. 2007. The ABC & XYZ of bee culture. 41st Edition. A. I. Root Co., Medina, OH. xxx
Simon, C. 2001. Neith. In Redford, D. B. The Oxford encyclopedia of ancient Egypt. Oxford University Press, New York. 2: 516.
Smith, W. S. 1940. News items from Egypt: the season of 1938 to 1939 in Egypt. American Journal of Archaeology 44(1): 145–149.
Sobhy, G. P. G. 1924. An Eighteenth Dynasty measure of capacity. Journal of Egyptian Archaeology 10(3/4): 283–284.
Sowada, K. N. 2009. Egypt in the eastern Mediterranean during the Old Kingdom: an archaeological perspective. Fribourg: Academic Press, Göttingen.
Spalinger, A. J. 2001. Festivals. In Redford, D. B. The Oxford encyclopedia of ancient Egypt. Oxford University Press, New York. 1: 521–525.
Spanel, D. B. 2001. Funerary figurines. In Redford, D. B. The Oxford encyclopedia of ancient Egypt. Oxford University Press, New York. 1: 567–570.
Strudwick, N. 2010. Hieroglyph detective. Chronicle Books, San Francisco, CA.
Sultanate of Oman. 2002. Beekeeping. http://www.omanet.om/english/culture/beekeep.asp?cat=cult.
Tyldesley, J. 2005. Egypt: how a lost civilization was rediscovered. University of California Press, Berkeley.
Velde, H. T. 2001. Seth. In Redford, D. B. The Oxford encyclopedia of ancient Egypt. Oxford University Press, New York. 3: 269–271.
Venable, S. L. 2011. Gold: a cultural encyclopedia. ABC-CLIO, Santa Barbara, CA.
Verner, M. 2003. The Fifth Dynasty's mysterious sun temples at Abusir. KMT 14(1): 44–57.
Vernus, P. 2003. Affairs and scandals in ancient Egypt. Cornell University Press, Ithaca, NY.
Vischak, D. 2001. Hathor. In Redford, D. B. The Oxford encyclopedia of ancient Egypt. Oxford University Press, New York. 2: 82–85.
Wente, E. F. 1966. The suppression of the high priest Amenhotep. Journal of Near Eastern Studies 25(2): 73–87.
Widmer, G. 1999. Emphasizing and non-emphasizing second tenses in the "Myth of the Sun's Eye," Journal of Egyptian Archaeology 85: 165–188.
Wiebach-Koepke, S. 2001. False door. In Redford, D. B. The Oxford encyclopedia of ancient Egypt. Oxford University Press, New York. 1: 498–501.
Wilkinson, R. H. 1992. Reading Egyptian Art. Thames and Hudson, London.
Wilkinson, R. H. 2008. Egyptian scarabs. Shire Egyptology, Oxford, UK.
Wilkinson, T. 2010. The rise and fall of ancient Egypt. Random House, New York.
Wilson, H. 1988. Egyptian food and drink. Shire Publications. Aylesbury, UK.
Wilson, H. 1988a. A recipe for offering loaves? Journal of Egyptian Archaeology 74: 214–217.
Zander, E. 1941. Beiträge zur Herkunftsbestimmung bei Honig, III Pollengestaltung und Herkunftsbestimmung bie Blütenhonig. Verlag Liedloff, Loth & Michaelis, Leipzig.
Zauzich, K. 1992. Hieroglyphs without mystery. University of Texas Press, Austin.
Zecchi, M. 1997. On the offering of honey in Graeco-Roman temples. Aegyptus 77 (1/2): 71–83.
Zenihiro, K. 2013. Data on funerary cones. https://sites.google.com/site/dataonfunerarycones/.
Zumla, A., and A. Lulat. 1989. Honey: a remedy rediscovered. Journal of the Royal Society of Medicine 82: 384–385.

INDEX

Printed in the USA/Agawam, MA
March 18, 2024

862918.006